THE HISTORY OF THE LASER

Other titles of interest

England's Leonardo: Robert Hooke and the Seventeenth-Century
Scientific Revolution
A Chapman

Masers and Lasers
M Bertolotti

A History of Light and Colour Measurement
S F Johnston

The Science of Imaging: An Introduction
G Saxby

Black Holes, Worm Holes and Time Machines
J S Al-Khalili

Superstrings and Other Things
C Calle

Minding the Heavens: The Story of our Discovery of the Milky Way
L Belkora

Origins: The Quest for Our Cosmic Roots
T Yulsman

THE HISTORY OF THE LASER

Mario Bertolotti

University of Rome 'La Sapienza'

Translated from *Storia del laser* by M Bertolotti
Bollati Boringhieri 1999

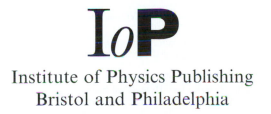

Institute of Physics Publishing
Bristol and Philadelphia

British Library Cataloguing-in-Publication Data
A catalogue record for this book is available from the British Library.

ISBN 0 7503 0911 3

Library of Congress Cataloging-in-Publication Data are available

Commissioning Editor: Tom Spicer
Production Editor: Simon Laurenson
Production Control: Sarah Plenty and Leah Fielding
Cover Design: Frédérique Swist
Marketing: Nicola Newey, Louise Higham and Ben Thomas

Published by Institute of Physics Publishing, wholly owned by The Institute of Physics, London

Institute of Physics Publishing, Dirac House, Temple Back, Bristol BS1 6BE, UK

US Office: Institute of Physics Publishing, Suite 929, The Public Ledger Building, 150 South Independence Mall West, Philadelphia, PA 19106, USA

Typeset by Academic + Technical, Bristol
Index by Indexing Specialists (UK) Ltd, Hove, East Sussex
Printed in the UK by MPG Books Ltd, Bodmin, Cornwall

CONTENTS

PREFACE

It is amazing how human imagination anticipated the invention of the laser. H G Wells in his famous novel 'The War of the Worlds' (1898) describes a death ray, and in the Flash Gordon comics (1950) light-ray guns are widely used; weapons which would be now identified as high-power lasers.

The word laser is by now well known to the layman, who is literally surrounded by applications of laser light, in the fields of medicine (surgical and diagnostic procedures), telecommunications (fibre-optic telephone links, compact-disk information storage and retriving, holograms) and technology (laser drilling of materials, geodesic measurements, newspaper printing).

Lasers come in different shapes, sizes and prices, and go under different names such as ruby (the first to operate), helium–neon, argon, semiconductor laser and others. Notwithstanding their popularity, very few people really know what a laser is and how it works. In this book, I will try to explain in the clearest possible way (although it will prove impossible to avoid some technical considerations) how people managed to build the first lasers and the principles by which they operate (together with the masers, their counterpart in the micro-wave range).

At this stage, it suffices to say that the laser is a light source with peculiar characteristics, drastically different from those of conventional sources such as a candle or a light bulb. In fact, laser light consists of a single colour (not a mixture of colours like white light) and is radiated in a single direction (not in all directions, as in a light bulb), which enables us to collect it with a lens and focus it in a region of very small dimensions. The spectral purity and directionality of laser light dramatically improves the efficiency of this procedure, making it possible to concentrate a sizeable amount of power in a small region for different operations, like the melting or cutting of a metal.

In the applications mentioned above, the laser is basically used as a very powerful light bulb. However, there are others (like optical communications), in which its most important characteristics are the spectral bandwidth and angular aperture of the emitted beam. To understand them, we need to consider what light is and how it is emitted, which in turn depends on the emitter, the atom, a task which requires the introduction of some basic concepts of quantum mechanics. We will discuss the different emission mechanisms, that is spontaneous emission—the dominating process in all natural sources—and stimulated emission—the process governing laser light and responsible for its peculiar characteristics.

In order to explain the different phenomena according to an historical sequence, we will retrace the story of light and the first steps of quantum mechanics. In so doing, we will appreciate that science is built gradually, like a jigsaw, and that many of its ideas, too advanced with respect to their historical context, are bound to remain unappreciated and unused, while others may blossom simultaneously and independently in the minds of many people, as if they were the unavoidable consequence of the preceding ideas and the indispensable premise of those to follow.

Before beginning our story I wish to thank The American Institute of Physics Emilio Segrè Visual Archives who provided the authorization to publish the photographs. A special acknowledgment goes to my wife who read the Italian text and suggested a number of changes which notably improved its clarity. Finally I wish to thank Tim Richardson for converting my Anglo–Italian into English, and Tom Spicer and Leah Fielding of Insitute of Physics Publishing for their encouragement.

INTRODUCTION

The creation of the world, as described in the first book of Genesis, is not actually at variance with the most recent cosmological theory of the Big Bang, according to which the Universe started with a great explosion of light.

But how does the light originate? A child would look at in astonishment and respond that light comes from the Sun or from an electric lamp or a fire. Certainly this would be correct. However, why does the Sun emit light and similarly, though to a lesser extent, why does fire? For thousands of years mankind did not ask, or rather linked light to philosophical and religious concepts, putting emphasis principally on the problems connected with vision, which in those early times were the most pertinent issues related to light. In Greek mythology we find the Titan Epimetheus, who assumed the task of giving to each animal of the Creation a particular characteristic to protect itself and survive. He provided the tortoise with a hard shell, the wasp with a sting, and so on, until when he came to the human race he had exhausted all the possibilities of nature and was unable to find anything for man. Plato writes that man stood 'nude, barefoot, without a house and unarmed'; Epimetheus asked for help from his brother Prometheus who stole fire from Zeus and presented it to man to help develop mankind, culture and technologies. Full of rage and jealousy, Zeus punished Prometheus by chaining him to the Caucasian mountains where every day an eagle tore at his liver. To prevent mankind enjoying the gift, he then ordered Ephesus to mould the first mortal woman, the beautiful Pandora, who married Epimetheus and through curiosity opened a box, given to her protection, full of all the evils of the world which spread and caused misfortune to all mankind.

In a similarly fantastic way the nature of light was clear to the ancient Egyptians, for whom it originated from the glance of Rah, their Sun god. A priest in 1300 BC wrote 'When the god Rah opens his eyes there is light; when he closes his eyes, night falls'.

It would be possible to quote many other examples showing that the problem of the origin and nature of light was in ancient times considered in a religious and fantastic frame.

1

The understanding of light of the ancient Greeks

Pythagoras, in the 6th century BC when in Greece philosophy and science were developing together, formulated a theory of light according to which rectilinear visual rays leave the eye and touch objects, so exciting visual sensation.

According to Empedocles (circa 483–423 BC), Aphrodite, the goddess of love, forged our eyes with the four elements with which he thought everything was made (soil, water, air and fire) and lit the fire, just like a man using a lantern to light up his path in the dark. Vision occurred from the eye to the object: the eyes emitted their own light.

Plato (circa 428–427 to 348–347 BC) assumed that the fire in the eye emitted light and that this interior light, mixed with daylight, formed a link between objects in the world and the soul, becoming the bridge through which the smallest movements of external objects generate visual sensation. According to the philosopher two forms of light—one internal and the other external—mix and act as mediator between man and a dark and cavernous external world.

The delicate beginnings of a transition towards a mechanical view of vision started with Euclid, the great Alexandrian mathematician who lived around 300 BC and in his writings on optics provided a clever geometric theory of vision. Euclid continued to believe that light came from the eye but, at variance with the vague luminous and ethereal emanation assumed by Empedocles and Plato, it became a rectilinear light ray to which mathematical deduction could be applied. In his extended mathematical studies the philosopher gave geometrical form to visual rays and developed some of the laws of geometrical optics as we know them today. He, and like him Archimedes (circa 287–212 BC) and Heron (3rd or 2nd century AD), joined Pythagoras and his disciples. Instead Democritus (470–360 BC) and the atomists assumed that the illuminated objects emitted atoms, which constituted images of those same objects, and which, when collected by the eye, generated vision.

The damage done by Aristotle

Later on, Aristotle (384–322 BC) defined light as 'the action of a transparent body, in that it is transparent' observing that a transparent body has the 'power' to transmit light, but does not become effectively transparent until light has gone through it and triggered its transparency.

If we observe the eyes of a cat at night, we notice they are bright and that cats can easily walk in the dark; this fact convinced ancient people of the real existence of a fire in the eyes as told by Empedocles and Plato. However, a prickly question arose: if a source of light exists in the eye, why is man not able to see at night? Answers were many, but Aristotle cut discussion short by insisting that dark air is opaque: only when a lamp is fired does it become transparent because light activates its latent transparency, after which man can see. We can still ask why the same reasoning did not apply to the cat that sees without the lamp fired. In any case all these considerations did not answer the questions concerning the nature of light and how it is produced. During the Middle Ages, when problems of nature were discussed on the basis of Aristotelian philosophy, according to which the 'nature' of things consists of the reasons for their existence, that is in their ultimate end, no progress was made to find a solution.

Saint Thomas Aquinas (1227–1274) declared that 'the origin of our knowledge is in the senses, even of those things that transcend sense' and 'metaphysics has received its name, that is beyond physics, because to us, who naturally arrive at the knowledge of things immaterial by means of things sensible, it offers itself by rights as an object of study after physics'.

Aristotelism was extensively adopted in 13th-century Europe, dominating for at least four centuries so much so that even in 1624 the parliament in Paris declared that, under sentence of death, nobody could support or teach doctrines opposed to those of Aristotle.

The scholars of the Middle Ages considered Aristotelism an encyclopaedic body of knowledge which could not be improved. They dropped the view of Saint Thomas concerning the relationship between physics and metaphysics affirming 'it is not the province of physics to theorize on its own facts and laws or to undertake a reconstruction of cosmology or metaphysics ... if a physical theory is inconsistent with received metaphysical teaching, it cannot be admitted, because metaphysics is the supreme natural science, not physics'. Accordingly they interpreted the external world by applying only formal logic, extracting deductions from dark and sterile principles which in reality represented the petrifaction of flawed Aristotelian physics, an approach which brought nothing more than a prolix sophistry which prevented scientific progress during the Middle Ages.

Although Aristotle may be considered one of the greatest philosophers—one of the founders of logic—his teaching arrived at the moment of decline of the creative period of Greek thinking, and instead of stimulating further intellectual activity it was accepted as a dogma and halted any other philosophical activity. Two thousand years later, at the time of the arousal of new philosophical thinking, practically any progress in science, in logic and in philosophy was forced to begin with an opposition to Aristotelian theories.

The rise of modern science

A necessary condition for the emergence of modern science was emancipation from the Thomist philosophy. The process was aided by a number of circumstances. During the 15th century, various causes contributed to the decline of the papacy which resulted in a very rapid political and cultural change to society. Gunpowder strengthened central government at the

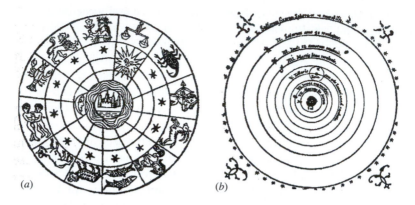

(a)　　　　　　　　　　(b)

Figure 1. (a) Ptolemaic model accepted up to the early 17th century: the Earth is in the centre and Sun and planets revolve around it. Planets describe small circles (epicycles) whose centres move on large circles (deferents) with the Earth as the centre. (b) Copernican vision of the solar system: the Sun is at the centre and the planets turn around it in concentric circular orbits (from Burgel B H 1946 *Dai mondi lontani*, Einaudi, Torino, p 37 and Abetti G 1949 *Storia dell'Astronomia*, Vallecchi, Firenze).

expense of the feudal noble society, and the new—essentially classic—culture venerated Greece and Rome and condemned the Middle Ages.

Decisive elements for the renewal of science were the new relationship between Earth and Sun as proposed by Nicolas Copernicus (1473–1543) in 1543, according to which the Earth revolves around the Sun and not vice-versa, as was assumed since the times of Ptolemy (Egyptian astronomer, mathematician and geographer, circa 100–178 AD) (see figure 1) and, at the beginning of the 17th century, the success of Kepler's (1571–1630) theories. Postulating three laws which rule the motion of planets around the Sun, Kepler demonstrated the falsity of the Aristotelian principle according to which celestial bodies are of a different species from terrestrial ones.

Kepler was born in the small town of Weil in Wurttemberg. He was educated to become a protestant pastor but, being in favour of Copernicus' ideas, was forced to give up this aspiration. His professor of mathematics and astronomy recommended him for a teaching position in Graz, where he published in 1596 his first work, *Mysterium Cosmographicum*, in which he clearly expresses his belief in a mathematical harmony of the Universe. Being a protestant, he was exiled when the Archduke Ferdinand began the rigorous counter-reformation, and took refuge in Prague on the invitation of the astronomer Tycho Brahe (1546–1601) with whom he collaborated until his death. He used the exact astronomical observations of Tycho to obtain his laws of planetary motion. After the death of the Emperor Rudolf II, he moved to Linz in an effort to defend, successfully, his mother from a charge of witchcraft. When in 1619 the Archduke Ferdinand was raised to the imperial throne with the name of Ferdinand II, the persecutions of protestants increased and in 1626 Kepler was forced to leave Linz. After travelling widely, he died in 1630 while journeying to Ratisbone to obtain justice from the Parliament. The Thirty Years War removed thereafter any trace of his burial, which was outside the city gates. Whilst not believing in them, Kepler, one of the architects of the astronomical revolution, produced horoscopes throughout his life to increase his meagre finances.

Kepler was fascinated by the old Pythagorean idea—which favoured the spherical form—and tried to find in the movements of planets the same proportions that appear in musical harmonics and in the shapes of regular polyhedra. In his vision, the planets were still living entities with an individual soul, like the Earth. The rejection of this fantastic view of the physical world, started by Galileo and ended by Newton, is barely alluded to by Kepler, and is present only in his scientific method of treating problems, at variance with the magic/symbolic attitude typical, for example, of alchemy.

The celestial bodies with the Sun at the centre are, for Kepler, a realization, although imperfect, of a spherical image of the Holy Trinity. Already in *Mysterium Cosmographicum* he writes: 'the image of the trine God is a spherical surface, i.e. the Father is the centre, the Son is the external surface and the Holy Spirit is like the rays that from the centre irradiate towards the spherical surface'.

From his examination of the motion of the planets he deduced that they revolve around the Sun, describing ellipses with the Sun at one of the foci (first law), and that the line which joins the Sun to a planet covers equal areas in equal times (second law). He then demonstrated that the orbits are not casual but the maximum distance of a planet from the Sun is in some ratio with the time employed to make a tour around the Sun itself (third law; figure 2).

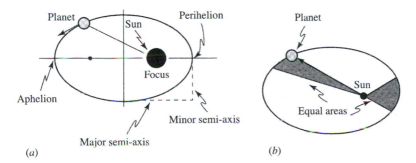

Figure 2. Kepler's laws. (a) The first law states that planets move on elliptic orbits with the Sun at one focus. The perihelion and aphelion are the point of minimum and maximum distance from Sun, respectively. (b) The second law states that the line which connects the Sun to the planet covers equal area in equal times. Therefore, the two shaded areas are equal if to cover each one the same time is employed, and the planet must have higher speed when travelling in the segment nearer to the Sun than when it travels in a more distant segment. The third law establishes that the square of the time a planet employs to make a full trip around the Sun is proportional to the cube of the major semi-axis of its orbit.

Descartes

René Descartes (Cartesius, 1596–1650), a young contemporary of Galileo, was the first to try to make a general reconstruction of the ideas concerning the physical Universe.

The period which preceded his birth and the one in which he lived were marked by events which greatly changed the conception of the world. The discovery of the Americas, the circumnavigation of the Earth by Magellan, the invention of the telescope, the fall of the Ptolemaic system in astronomy, and the general dissatisfaction of scholasticism contributed to weaken the old foundations and provide the basis for a new structure.

> Descartes is considered the founder of modern philosophy and one of the originators of 17th century science. His father, Joachim, counsellor of the Parliament of Brittany, was a moderate land-owner and his mother, who died when he was still a child, left revenue which allowed him to live. The young Descartes followed the military profession, and served in the campaigns of Maurice of Nassau and of the Duke of Bavaria. Being 24 years old, after a deep mental breakdown, he decided to devote himself to philosophy.
>
> He was gifted with a strong and brilliant imagination which made him a single-minded man both in his private life and in his way of reasoning. Voltaire says of him, not without irony, that 'nature made him nearly a poet and in effect he composed for the Queen of Sweden a *divertissement* in verse that in the honour of his memory was not printed'. Believing that to develop philosophy freely, it was necessary to escape fellow men and chiefly his fatherland, he took refuge in Holland and, after a period during which his ideas were opposed both there and in France, his genius was eventually recognized. France called him back, promising him a brilliant position which in the end he did not receive. So he accepted an invitation to Stockholm from Queen Christine of Sweden, but being obliged by the passion that the Queen had for philosophy to give her private lessons in the early hours of the icy Swedish mornings, the philosopher, who was used to resting in the warmth of his bed, died of pneumonia after only a few months.

Descartes' objective was to create a theory of the Universe worked out as far as possible in every detail. Such theory was necessarily forced to have a metaphysical basis, and in fact metaphysics is the most well-known part of his work. The first step was to discard the futile methods of the Middle Ages, the trials to interpret nature in terms of act and power, matter and form, substance and accident, the ten categories and similar Aristotelian concepts, as already proposed by other philosophers. In fact already Francis Bacon (1561–1626) and Galileo Galilei (1564–1642) had started the transformation that led to the rejection of Aristotelian rule that in physics attributed all to the final cause. Galileo founded modern science, by introducing the experimental method, and based it both on the necessity to observe the external world and to question nature through suitable experiments.

Galileo

Galileo Galilei was born in Pisa in 1564 and following the wish of his father, a scholar and qualified musician, entered Pisa University in 1581 to become a physician. Medical studies, however, interested him little, whilst he was greatly attracted by mathematics. In 1583 he made his first important discovery by observing that the lamps which hanged from the Pisa cathedral roof oscillated under the influence of the wind, yielding always the same time period to perform a full oscillation, irrespective of its amplitude.

After a number of unsuccessful attempts, in 1589 he obtained a chair of mathematics in Pisa, a secondary teaching post with a modest annual salary of 60 scudi. Essentially for economical reasons, he moved to Padua in 1592. His still modest revenue obliged him to give private lessons, and to produce mechanical instruments in a small workshop together with a technician. In Padua, in 1609, being informed that the previous year a Dutch scholar had invented the telescope, he was able to assemble one, and looking at the Moon saw immediately it was not the uniform and smooth object speculated by Aristotle. He also discovered the sunspots, but the most exciting moment was when he discovered four of Jupiter's satellites. They were new bodies, not quoted by Aristotle and certainly did not rotate around the earth. He immediately published his observations in *Sidereus Nuncius* (1610).

At the end of 1610 he came back to Pisa where he studied several problems concerning motion, which he later partly published in *Dialogo sopra i due massimi sistemi del mondo* (1632) whose content is strictly astronomical and where a 'Socratic' confutation of the old physics and cosmology (defended by Simplicius) is made by the Copernican protagonist Salviati. All his mechanical observations were later collected in *Discorsi e dimostrazioni matematiche intorno a due nuove scienze attenenti alla meccanica e ai movimenti locali* (1638).

Galileo discovered the importance of acceleration, that is the change of speed with respect to time, both in magnitude and direction. Until that moment, people thought the motion of a terrestrial body would gradually cease if left to itself.

Why do bodies move? For nearly two thousand years, on the basis of the false assumptions made by Aristotle, everybody thought a force was necessary to maintain a body in motion. This belief appeared wise considering that, for example if a horse gives up pulling a cart, the cart stops quickly. Aristotle's followers extended their reasoning to the motion of projectiles by maintaining that an arrow could continue to move because the air flew from tip to tail pushing it forward; this is a beautiful example of Aristotelian logic.

Really even the question 'what makes a body move?' is misleading. This question made sense for Aristotle, who believed that the natural status of a body is rest. Galileo performed a series of experiments that convinced him that a moving body possesses a quantity of motion (momentum) that conserves itself. He maintained that any body left to itself maintains its motion along a straight line with uniform velocity. The meaningful question is therefore not 'what makes a body move?' but 'what changes the motion of a body?' Galileo answered that any change in the speed or direction of motion must occur due to the action of some force. Such a principle was later expressed by Newton as the first law of motion, or Law of Inertia, and today we can easily verify it by making a disc glide along a plane. The smoother the plane, the farther the disc continues in its motion.

We may also make experiments using objects which move on an air cushion in a special apparatus allowing us to approach the limit of zero friction, and verify thereby that even a small kick makes the object glide with a small but mostly constant velocity. By extrapolating this experimental result, it is then possible to say that if friction could be completely eliminated, the body would continue to move indefinitely with constant velocity. An external force is therefore needed to put a body into motion but, once moving, no force is necessary to allow it to travel with constant velocity.

Further still, Galileo expressed the law governing the motion of a falling body, maintaining that when a body falls freely, its acceleration, if the air resistance is neglected, remains constant and is the same for all bodies. In fact he demonstrated experimentally that a weight of lead falls with the same rapidity as a smaller mass. He also asked in his *Discorsi intorno a due nuove scienze* if light propagates in an instantaneous way or with a finite speed, and devised an experiment which may be considered the forerunner of a later experiment in 1849 by A H L Fizeau (1819–1896).

He warmly adopted the heliocentric system; he was in correspondence with Kepler and accepted his discoveries. With his telescope, he discovered that the Milky Way is formed by a great number of stars and was able to observe the phases of Venus, which were implicit in Copernicus' theory but could not be observed with the naked eye. When on 7 January 1610 he discovered the four satellites of Jupiter that, in honour of his protector Cosimo II de' Medici, he called 'stelle medicee' he came across a small problem: everybody was aware of the existence of seven celestial bodies, that is the five planets, Mercury, Venus, Mars, Jupiter and Saturn, the Sun and the Moon. Seven was considered a sacred number: is not Saturday the seventh day? Did there not exist the seven armed candelabrum and the seven churches in Rome? Were there not seven capital sins? What could be more proper than the existence of seven celestial bodies? If we add now the four moons of Jupiter, the seven celestial bodies become eleven, and this number has no mystical property. Therefore traditionalists refused to accept the telescope, they refused to look through it, and maintained it showed only fantasies. Galileo laughed together with Kepler about this, but soon stopped when the Inquisition condemned him to prison and obliged him to deny the motion of the Earth. Science in Italy did not progress for centuries but the Inquisition was not able to prevent men of science adopting the heliocentric theory, especially in the protestant countries.

In his last years Galileo was allowed to return to his house in Arcetri where he died as a blind man in 1642. A solemn burial in the church of Santa Croce in Florence was forbidden by Rome because it could 'scandalize wise men' and 'give offence to the reputation' of the Inquisition. The prohibition to read Galileo's works was removed only in 1757. A partial rehabilitation was granted by Pope John Paul II in 1981 with the nomination of a pontifical commission, divided into four study groups which were charged with analysing the exegetic, cultural, scientific and historical-juridical aspects of the trial. In October 1992 the commission reached a final conclusion not by rehabilitating Galileo but acknowledging that all protagonists of a process, with no exception, have the right to the benefit of bona-fide in the absence of unfavourable documents. This was a rather vague way of saying that at the time the trial could have been conducted differently.

The physics of Descartes

Kepler's discovery of the three laws of the motion of the planets pointed out the supreme importance of mathematics in the study of nature, and gave the principal inspiration to Descartes whose research was dominated by the conviction that the theorems of mathematics gave a precision, a certainty and a universal acceptance not found in other disciplines. As a consequence Descartes based all his constructions on the axiom that clarity and certainty are the marks of all genuine knowledge. He started by denying that a body could act on another being far apart, affirming that bodies may interact only when they are in contact. As a consequence the space between the Moon and the Earth and, more generally, the whole of space could not be empty, but was filled partly by ordinary bodies, like air and tangible objects. The interstices among the particles composing these bodies and the remaining space was assumed to be 'a plenum' filled by a medium which although imperceptible by human senses, was able to transmit a force and exert effects on material bodies plunged into it. This plenum he called 'aether'. So the term ether lost its meaning given in ancient Greek cosmology, of an incorruptible element constituting celestial spheres and bodies. Descartes' particles are in continuous motion forming vortices, and light is simply the transmission of a pressure exerted on the eye from the motion of the vortices. In his work *Dioptrique*, the scientist compared vision to the perception of an object detected by a blind man using his stick.

Based on the idea that the effects produced by contact and collisions were the simplest and most understandable phenomena of the external world, he did not admit any other agent. He did not require, as we do now, that his scheme had an experimental verification, because he trusted more in the simplicity and precision of speculation than in its correspondence with the observed fact.

His work must be considered a stupendous mental effort intended to show that the whole Universe and its origin can be represented through a logically coordinated mechanical scheme which depends only on a few very basic physical actions, and that a complete understanding of its working can be found with the use of mathematics. He originated the idea of mechanical philosophy according to which the external inanimate world may for scientific purposes be regarded as an automatic mechanism, and a mechanical model may be conceived for every physical phenomenon.

A similar point of view could not be conceived before the Renaissance when there was only very little experience—if any at all—of the self-governing mechanisms which operate independently from human will. People were aware only of tools that to accomplish some work required intelligent guidance, and therefore any manifestation of regularity was understood to be the result of the activity of some mind. Already the Greeks believed that the order and harmony observed in the motion of celestial bodies relied on their soul, and many phenomena attracted absurd explanations in Aristotle's philosophy. For example, the falling of a heavy body was explained by assuming that weighted matter tended to seek its natural place: the centre of the Universe. This explanation became unsatisfactory when the Copernican theory of the solar system was accepted, because now the Earth was moving in infinite space and no point could be identified as the centre of the Universe. In a revolutionary way, Descartes suggested that the cosmos could be considered an immense machine and consequently all happenings in the material world could be predicted by mathematical calculations.

He went further, maintaining that physics, like Euclidean geometry, may be entirely derived from *a priori* principles, without any dependence on observation and experiment so supporting the doctrine of epistemological rationalism. In this belief he departed from the new doctrines of Bacon and Galileo and was criticized even by Huygens.

Really by building up an all-comprehensive theory of the Universe without having studied in detail any process, Descartes continued the Greek tradition rather than the new path traced by Tycho Brahe, Kepler and principally by Galileo. He never grasped the principle that true knowledge can only be gained gradually through a patient interrogation of nature, and his hypothesis that force could only be communicated through a pressure or impact did not allow him to explain any of the forces existing in nature. The defects of his methodology led in less than a hundred years to the rejection of nearly all of his theories; however, his ideas stimulated scientific thought at the highest level.

The law of refraction

In the work *Dioptrique*, Descartes presents his theory of light based on vortices, and discusses the laws of reflection and refraction, expressing for

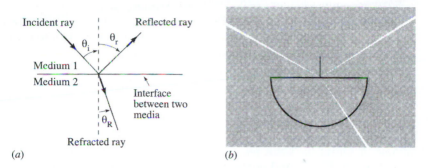

Figure 3. (a). Incident, reflected and refracted rays. The law of refraction states that $\sin\theta_i/\sin\theta_R = v_1/v_2$ where v_1 and v_2 are the velocity of light in the first and second medium, respectively. (b) A light beam reflected and refracted by a piece of glass.

the first time the principle that the ratio of the sines of the angles of incidence and refraction depends on the medium through which the light travels.

Already in the Greek world, people knew that if a light ray travels from one medium to another, it is partly reflected and partly transmitted at the surface of separation of the two media (see figure 3). Anyone may perform the experiment with a piece of glass exposed to a light ray coming from the window in a room. Part of the sunlight is reflected from the glass surface creating a light spot which moves around the walls by moving the glass, and part travels through the glass. The phenomenon by which a portion of the light goes through the glass is called refraction. The term comes from the Latin *refractio*, and originates from the fact that an object that is partly in one medium and partly in another medium (for example a stick that is partly in air and partly in the water) appears to be broken (in Latin *refractus*).

There are three laws of geometrical optics: the first affirms that light propagates in a straight line if it does not encounter obstacles; the laws of reflection and refraction are the second and third ones. The first law may already be found in the work on optics ascribed to Euclid (300 BC) which also contains the law of reflection; but both laws were known before.

The phenomenon of refraction of light was surely known by Aristotle. Later on, Ptolemy tried, without success, to establish a quantitative law. The measurements made by him related to relatively small angles, and he reached the wrong conclusion that the angle of refraction was proportional to the angle of incidence. Much later, the Arabian optician Alhazen (Abu Ali al-Hasan Ibn al Haitham 965–1038) found that the ratio between the incidence and refraction angles does not remain constant by changing the incidence angle, but he was not able to give a correct formulation of the law. To him are attributed around one hundred publications; in the most important one, translated into Latin in the 12th century and published in 1572 under the title *Opticae Thesaurus*, the Greek theory according to

11

which the eye emits rays was authoritatively rejected for the first time.

The correct formulation of the law of refraction is generally attributed to Willebrord Snell (1591–1676), a mathematics professor in Leyden, who around 1620 established experimentally the constant ratio between the cosecants of the incidence and refraction angles respectively, exposing these results in a manuscript that received some diffusion. Further historical research has shown, however, that the law of refraction was discovered by the English astronomer and mathematician Thomas Herriot (1560–1621) at Isleworth (Middlesex), around 1601. However, already the mathematician Abu Said al-Ala in his book *On the Burning Instruments* (written in about 984) had stated the law and computed with it a burning glass. Descartes in his *Dioptrique* contributed the modern formulation of the law affirming that the ratio between the sines of the incidence and refraction angles respectively is equal to the light velocity in the second medium divided by the velocity in the first one (i.e. in the medium from which the light arrives). By using his theory of light that, by analogy, he now assimilated to small projectiles, he demonstrated that the law was due to the change of propagation speed of the projectiles travelling from one medium to the other. According to Huygens, however, he had read Snell's manuscript and taken advantage of it.

Although it is true that the ratio between the sines of the incidence and refraction angles depends on the speed of light in the two media, Descartes, by applying the laws of mechanics to small projectiles by which he represented the light, found that to explain the experimental circumstance that in the denser medium the ray inclines less, one must assume that in it the light particles travel faster, whilst in reality the opposite is true.

In any case the theory had great successes. He provided the mathematical explanation of the rainbow, by calculating the reflections and refractions of light in rain droplets, and this was considered a divine result by his contemporaries. Until a few years before, in fact, the rainbow was looked at as an inexplicable miracle and it was the philosopher and Archbishop of Split, Marco Antonio De Dominis (1560–1624), who guessed that it is a phenomenon connected with the rain and the sun.

Descartes' light theory displaced quickly the medieval views, stimulating new studies. Attempts to explain rationally the nature and origin of light triggered, however, a great controversy and two contrasting theories developed: the wave theory by Hooke and Huygens and the corpuscular theory introduced by Descartes and continued by Newton and his supporters.

CHAPTER 1

WAVE AND CORPUSCULAR THEORIES OF LIGHT

The three characters who played a central role in the history of the theory of light were Hooke, Huygens and Newton. Hooke and Newton were from Britain; Huygens was Dutch. They made notable contributions to several fields of physics, and established the basis for a modern understanding of light even though they proposed contradictory theories. One was based on a wave interpretation, while the other considered light to be composed of small particles. The two theories, which seemed irreconcilable, aroused fierce discussion and argument among both the protagonists and their supporters. A torrent of words has been written about this dispute; here we will not go deeply into the subject but limit ourselves to the most important facts.

Robert Hooke

Robert Hooke was born in Freshwater on the Isle of Wight in 1635 and died in London in 1703. He was a single-minded man who designed and built a number of instruments and devices: we owe to him, for example, the innovative use of the spiral spring in the balance mechanism of clocks, which allows their accurate operation. He expressed the law of proportionality between elastic deformations and strain which bears his name, performed a number of astronomical observations, and vindicated some of the work of Newton regarding the discovery of the Law of Gravitation, of which he really had only a vague intuition.

Through meticulous observations with a 20 m long telescope that vibrated in the slightest breeze, he was the first to describe the shadow that Saturn's ring cast on the planet and to make detailed maps of the Moon's craters. He was an accomplished surveyor and architect who helped to rebuild London after the Great Fire of 1666. He was one of the first to articulate the concept of extinction and who suggested evolution two centuries before Charles Darwin.

He started his scientific career as an assistant of the well-known English chemist Robert Boyle (1627–1691) who, after studying at Eton and Geneva, in Oxford in 1654 began work on the properties of gases and vacuums.

In 1658–59, Hooke built a vacuum pump for Boyle of the kind invented in 1650 by Otto von Guericke (1602–1686), and helped in the demonstrations Boyle gave in those years during which he showed that air is essential for the transmission of sound, breathing and combustion. In the same period, Boyle expressed the famous law—which in the UK bears his name and in continental Europe is often attributed to Mariotte—according to which the product of the pressure of a gas and its volume is a constant if, just as pointed out by Mariotte, temperature remains constant. Boyle was so satisfied with Hooke that in 1662 he appointed him first Curator of Experiments at the Royal Society for the Improvement of Natural Knowledge, the English Academy of Sciences, founded by him in 1660 together with other English scientists under the auspices of Charles II. The role of the society was 'to improve the knowledge of natural things' by observation and experiment rather than by appealing to literary authorities: the society's motto is *Nullius in Verba* ('on the words of no one'). To this end the society appointed a curator of experiments. In this role, Hooke, whenever the Society held its meetings, was required to perform three or four experiments, and even if the Society hosted a meeting only once per week, the task was rather onerous and binding. However, he performed his duty with great dexterity for many years until he was appointed Secretary of the Society.

In optics he made two important experimental observations, both of which, however, were anticipated by others. The first discovery, that he recounted in his *Micrographia*, published in 1667, detailed the observation of the iridescent colours that can be seen when light falls on a thin layer of air between two glass plates or lenses, or on a film of any transparent substance; the so-called colours of thin plates or Newton rings, already observed by Boyle and later studied fully by Newton.

Hooke's second finding, made after the publication of his *Micrographia*, was that light in air does not propagate exactly along a straight line but some illumination can be found also in the geometrical shadow region of an opaque body. This phenomenon had been already observed and published in 1665 in a posthumous work by Francesco Maria Grimaldi (1613–1663), an Italian Jesuit who gave the phenomenon the name 'diffraction'.

Hooke's theoretical investigations of the nature of light have acquired notable importance, because they represent the transition from the Cartesian system to a complete wave theory. At variance with Descartes, Hooke concluded that the condition associated with light emission from a luminous body is a rapid vibratory motion of very small amplitude. On the subject of light propagation he wrote:

> '... in an *Homogeneous medium* this motion is propagated every way with *equal velocity*, whence necessarily every *pulse* or *vibration* of the luminous body will generate a Sphere, which will continually increase, and grow bigger, just after the same manner (though indefinitely

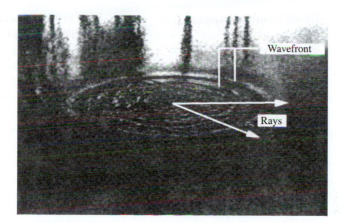

Figure 4. Waves on the surface of a lake. The circular ripples represent the wave fronts. The rays, perpendicular to the wave fronts, show the wave motion direction.

swifter) as the waves or rings on the surface of the water do swell into bigger and bigger circles about a point of it, where by the sinking of a Stone the motion was begun, whence it necessarily follows, that all the parts of these Spheres undulated through an *Homogeneous medium* cut the Rays at right angles.'

Hooke introduced also the concept of the 'wave front' or locus of points for which, at each time, a disturbance (that is any variable taken to define the perturbation), originally generated at a point source, has the same value. He affirmed that it is a sphere whose centre is the emission point, and whose radii correspond to the light rays issuing from the source (figure 4).

Hooke reconsidered Descartes' theory of diffraction, and deduced the law of diffraction from the deflection of the wave front. He assumed that the deflection of the wave front was responsible for the prismatic colours that can be observed when a white light beam passes through a glass prism. He supposed that white light was the simplest type of disturbance, made by a simple and uniform pulse at right angles to the propagation direction, and assumed colour was generated by the distortion of this disturbance during the refraction process. We will see that this theory was completely superseded a few years after its publication.

Christiaan Huygens

Christiaan Huygens, one of the founders of mechanics and physical optics, was born at Den Hague in 1629. He was a son of Constantijn (1596–1687), a well known poet of the Renaissance. Initially he studied rhetoric and law but, being a lover of sciences, he changed to mathematical studies. In 1655, by means of a powerful homemade telescope, Huygens clarified the problem of the configuration of Saturn by discovering its rings. He also discovered one of Saturn's satellites: Titanus (Titan). Further he recognized that the Moon is deprived of any atmosphere and its 'seas' must be empty of water. A year later, he wrote the first textbook on probability calculus, *De ratiociniis in ludo aleae*, and the following year built the pendulum clock.

In 1665 on the invitation of Colbert, the powerful state secretary of Louis XIV, Huygens settled in Paris, where in 1666 he was elected a member of the newly established Académie Royal des Sciences, and where in 1673 he published the *Horologium oscillatorium*. The book introduced the notion of moment of inertia, the first theorems on the mechanics of rigid bodies and the theory of the compound pendulum. By means of these investigations and others previously performed on collision phenomena, he expressed—albeit for a particular case—the theorem of work and energy, examined circular motion, gave the fundamental theorems on centrifugal force and verified that acceleration due to gravity changes as a function of latitude, continuing in this respect the research of G Borelli (1608–1679) and paving the way for Isaac Newton.

As a consequence of the savage campaign made by Louis XIV against Holland, in 1681, Huygens left France returning to his fatherland, where he and his brother devoted themselves to the construction of lenses for telescopes. In 1690 he published in Leyden his famous *Traité de la Lumière* in which, at variance with the emissive theory of Newton, he supported the wave theory of light. He died at Den Hague in 1695.

Huygens agreed with Hooke that light was essentially a form of motion. One had to establish if the motion was that of a medium or if it could be compared with that of a flight of arrows, as in the corpuscular theory, and he decided that the first alternative was the only valid one.

Evangelista Torricelli (1608–1647)—the scholar who after Galileo was appointed to the place of mathematician and philosopher of the Granducato di Toscana—had shown that light propagates very rapidly both in vacuum and in air, and Huygens inferred that the medium in which light propagates should permeate all matter being present in the whole universe even in the so called vacuum. This medium he called 'aether'. Accordingly light is a disturbance of the aether consisting of elastic vibrations that propagate at high speed in this highly elastic medium made of very tenuous matter. In 1675 the Dutchman Olaf Roemer (1644–1710) had already measured the velocity of light through astronomical observations, finding a value of $214\,300\ \text{km s}^{-1}$, which is about a third lower than the accepted value today of around $300\,000\ \text{km s}^{-1}$.

Huygens invoked the point of view introduced by Descartes, according to which every phenomenon can be represented by a mechanical process. Elastic phenomena were well known at the time, thanks also to Hooke's investigations, and if light needed to have a wave representation, elastic vibrations of some medium seemed appropriate. A number of examples of elastic vibrations were already known: sea waves are a wave phenomenon consisting of the oscillation of water; sound waves are an undulatory phenomenon of air, the sound emitted by a violin cord originates from its vibration. All these waves are elastic vibrations of some medium. Once it was established that the aether was the medium in which light propagated and that light was a wave, it was natural to represent it just as an elastic vibration of the aether.

In discussing the wave process of propagation, Huygens in his *Traité de la Lumière* introduced the famous principle that bears his name, with which he was able to quantitatively derive the reflection and refraction laws. Then, he gave a physical explanation of the change in the velocity of light as it propagates from one medium to another, by assuming that transparent bodies consist of hard particles that interact with aether, modifying its elasticity.

Both Hooke and Huygens assumed that light is a fast vibratory motion of the aether, the aether being defined as a particular medium, elastic and necessarily solid yet very soft (largely incompatible properties) which fills the whole space within and outside material bodies. This motion could be likened, as Huygens wrote, to the waves that can be observed on water on throwing a small stone.

Isaac Newton

Isaac Newton was born in Woolsthorpe, Lincolnshire, on 25 December 1642. He may be considered one of the truly great investigators of nature. His mother belonged to moderately noble classes and his father, who died before his birth, was a small landowner. His rigid puritan education, the enforced detachment from his mother who remarried to a protestant bishop and the successive rebuilding of a close affectionate relationship with her, encouraged some biographers to suggest Freudian interpretations concerning his neurotic stress, misogyny and temper whims.

He was destined to become a country squire, and undertook his first studies at the Grammar School of Grantham. His notable skills for mechanical invention and his inclination for humanistic studies in Hebrew and theology inspired his teachers to recommend him to Cambridge University. He was admitted to Trinity College in June 1661 but, due to the avarice of his mother, was forced to accept the title of *subsizar*, a term used in the University to mean a poor student who earned his keep by performing menial tasks for his fellow students. In this position, as had occurred previously at Grammar School, Newton was isolated from his fellow students who, when he subsequently became famous did not remember meeting him. He did not distinguish himself in his official studies, and in addition he did not follow an orthodox curriculum. Both in mathematics and in natural philosophy (that is to say in physics) he was a self-taught man, since at that time in Cambridge both disciplines were taught very little. The mechanical world of Descartes and the atomistic conception of the theologian, mathematician and astronomer Pierre Gassendi (1592–1655) who was a professor at the Collège de France in Paris, fascinated him, as he wrote in his notebooks written in that period under the title *Quaestiones quaedam philosophicae*. Although he did not seem to have reached any conclusion, clearly he was inclined towards atomism. He criticized Descartes' light theory, based on vortices, favouring a corpuscular theory. With natural philosophy he discovered mathematics. It seems that having bought the *Geometry* of Euclid, he read only a few pages, finding it obvious, although later he regretted not having paid more attention to the text.

To remain in Cambridge, Newton needed to win a permanent position and, maybe with the help of some sponsor, he was successfully elected to a scholarship in 1664. Thanks to his new status, he secured himself four years of study, and the liberalism of the school allowed him to completely devote himself to his studies. When he was working on a problem, he forgot not only to sleep but even to eat, so much that his cat became fat eating the food he left, and his contemporaries, who ate plentifully, remained astonished at his behaviour.

At that time, Cambridge was a university full of people looking only for financial security who had no great interest in studying; therefore the young Newton could easily cover the whole academic curriculum: in 1665 he earned his bachelor of arts degree, in 1667 was appointed minor fellow and in 1668 major fellow and master of art. In 1669 his teacher, the theologian, Hellenist, and mathematician Isaac Barrow (1630–1677) gave him the chair of mathematics (at present the reason is still unclear, it being barely credible that he did so only because he was struck by the genius of his pupil) which was held by Newton until 1701. This chair was financed by Henry Lucas who in 1664 left part of his fortune to establish a mathematics chair at Cambridge University, the first one after the five chairs established by Henry VIII in 1540.

Between 1664 and 1665 Newton became the most able mathematician of his time by establishing the basis of infinitesimal calculus and reaching important results in the study of mathematics.

At only 27, he was mathematics professor at Cambridge and somewhat later a member of the Royal Society. Tradition describes him as a distracted professor, continually engrossed in difficult problems, and recounts that, during the Great Plague in 1666, whilst taking refuge in the country at Woolsthorpe and walking in the garden, he observed an apple falling and tried to understand why the Moon does not fall to Earth. Following this train of thought led him to the discovery of the Law of Gravitation.

His rapidly-moving academic career was sustained by a solid foundation both in physics and mathematics, acquired during his solitary studies, as well as in the humanistic disciplines taught directly at Cambridge. His note-books contain precise recollections of his reading of Galileo, Descartes, Gassendi and more. The young 'natural philosopher' found himself raised 'on the shoulders of giants' as he said once, and was able to digest all the advances of the new science and develop with an extreme lucidity the principal themes of his great contributions as early as the years 1665–66, a particularly fruitful period of creativity and maturation.

By studying Kepler, Descartes and Galileo, Newton found himself considering the still unsolved problems of Copernican astronomy, and to reflect on atomism, on vacuum, and on the experimental and mathematical methods of Galileo. He put the problem of the planets' orbits—defined by the three laws of Kepler—in a new theoretical context, abandoning Kepler's hypothesis concerning the causes of celestial motions. At that time he gave the first formulation of the Law of Gravity, later to become legendary in the famous anecdote of the apple, which his biographers divulged and Newton loved to tell in his old age. In reality, recent studies show that the formulation of the Law of Gravity was made later. Questioned on how he discovered this law, he answered: 'by thinking at it continuously' and no other answer could better characterize the man, not only to underline his life dedicated to speculation rather than action, but also to understand his working methodology.

At that time he assembled the first reflective telescope, melting and grinding personally lens and mirrors. He also solved the problem of the composition of white light, through the famous experiments on dispersion performed by a glass prism. Those experiments were common among the scholars of optics, but only Newton was able to treat them with a precise mathematical method. In 1666, when he was still a student at Trinity College in Cambridge, he held a glass prism 'to try therewith the celebrated Phaeno-mena of Colours' (figure 5). As he wrote:

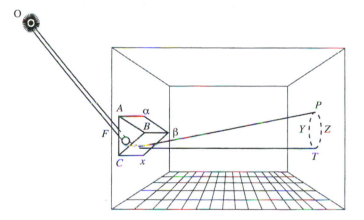

Figure 5.   This is the Newton's representation of the prism's experiment. A sun beam *OF* after passing through the small circular hole *F* is refracted by a prism *AαBβCχ* and is displayed in the spectrum *PYTZ* on the front wall. (From Add. Ms. 4002, p 3, Cambridge University Library.)

19

'having darkened my chamber, and made a small hole in my window-shuts, to let in a convenient quantity of the Sun's light, I placed my Prisme at his entrance, that it might be thereby refracted to the opposite wall. It was at first a very pleasing divertisement, to view the vivid and intense colours produced thereby; but after a while applying myself to consider them more circumspectly, I became surprised to see them in an oblong form, which, according to the received laws of Refraction, I expected should have been circular'.

The thin ray of sunlight which came through the circular hole made by Newton had in fact a circular cross-section and according to Snell's Law it should have changed only its direction but not its shape.

Newton and light

Newton tells us that having started the study of this strange phenomenon and calculating the ratios existing between the angle of incidence of the white light entering the prism, and the refraction angles of the coloured rays coming out, he immediately realized that each colour corresponded to a different ratio. By repeating the experiment and introducing a second prism, he noticed that each one of the fundamental colours maintains its ratio: he concluded that white light is 'an undefined aggregate of variously coloured rays'.

By describing this experiment, Newton perpetrated a mystification; the crucial experiment (*experimentum crucis* as Newton called it) on which he put so much emphasis, was largely a forgery invented later to explain his reasoning. Actually we know that Newton arrived at his result through a much more complex path, that we do not need to retrace here.

In conclusion we may agree with Newton that the angle of refraction depends on the colour of light, and that the white light of the Sun is so because it contains all colours, and is decomposed into its different colours by the refraction caused by a prism. Because the angle of refraction, as we said earlier, depends on the light propagation velocity, we may also say that the result obtained by Newton is that the propagation velocity depends on the colour of the light.

Today, the phenomenon according to which the light propagation velocity depends on its colour has been given the name of 'light dispersion', and 'dispersive power' is described as the ability of a given material to disperse the light of various colours on an observation screen using a prism. The reason why light of each colour propagates in the same medium with different velocity remained a mystery until the beginning of the 20th century.

Overcoming his unnatural reluctance to publish his discoveries, eventually Newton exposed his conclusions in 1672, by sending a letter to the Royal Society, and we believe it is interesting to remember how this happened. The theory of colours disclosed in the *Micrographia* by Robert Hooke did not satisfy Newton. Hooke maintained that 'blue is the impression on the retina of an oblique and confused pulse whose weaker part precedes and the stronger part follows' and that red is the impression of an oblique and confused pulse of inverted order. Newton in his *Quaestiones* contested immediately these two fundamental assumptions by Hooke, and performed the experiments we have just described showing that white light is a confused sum of coloured lights which are spread by the diffraction of the prism. Later he recognized that this result had an important practical consequence. With the diffusion of the telescope, experiments demonstrated that spherical lenses did not refract parallel rays, such as those coming from stars, in a perfect focus. In his work *Dioptrique*, Descartes had shown that lenses of an elliptic or hyperbolic shape due to the Law of Refraction were able to refract parallel rays into a perfect focus (i.e. exactly to a point). Newton started to study how to make similar surfaces and realized that because blue rays are diffracted more than red ones, he had obtained coloured images (today this defect is named 'chromatic aberration' of lenses). So he interrupted his work on non-spherical lenses and never resumed it, deciding to build by himself a reflective telescope, made in such way that the magnification of the image was obtained using a concave mirror instead of the lenses used in the Galilean telescope. The Newtonian telescope had a magnification of 40. Later he built a second one with a magnification of 150 and exhibited it in Cambridge. The news arrived at the Royal Society, who asked to see it and, by the end of 1671, received it from Barrow. The telescope produced a great sensation, so much so that in January 1672 Henry Oldenburg, the Secretary of the Royal Society, wrote to Newton praising his invention and announcing that the Society wanted to send a drawing to Huygens to prevent other people appropriating the idea, and asking permission to do so. Newton agreed, and on 6 February sent to the Society an account of his theory of colours explaining how this theory had brought him to the design of the reflective telescope.

Initially he had not formulated any hypothesis on the theory of light, but later on, after criticism by Hooke, Huygens and other people, to which he replied vigorously, he was forced to take a stance. In a memoire, appearing in the *Royal Society Philosophical Transactions*, he granted his adversaries a corpuscular hypothesis concerning the nature of light, not excluding, however, a wave alternative. The harsh controversy which arose with Hooke, which lasted for many years, induced him to publish nothing more on optics for a long time. His work *Opticks*, in fact, was published only in 1704, after the death of Hooke.

In the meantime, when he was appointed Lucasian professor at Cambridge in the autumn of 1669, Newton chose to deliver his inaugural series of lectures, given between 1670 and 1672, on the theory of colours and refraction which he had developed in the previous five years. These *Lectiones opticae*, written in Latin, were the first physical treatise and the most comprehensive account of his theory of colour. They were used as a basis for the first book of his *Opticks*, written 20 years later. By contrasting the *Lectiones* with the *Opticks* we may see the evolution of Newton's thought. In his *Lectiones*, Newton tried to create a mathematical science of colours, whilst as he himself declared, *Opticks* is an experimental textbook: 'My design in this Book is not to explain the properties of light by hypotheses but to propose and prove them by reason and experiment'.

The great Newtonian physical revolution
Later, in 1679, Newton resumed the study of bodies subjected to gravitational forces, solving the problem completely. The intuition he had in 1666, in fact, had not been fully developed because he had no precise measurements of the Earth from which to compute the attraction between Earth and Moon. After the Frenchman Jean Picard (1620–1682) measured the length of the meridian between Amiens and Malvoisine (1669–70), so allowing a precise estimation of the radius of the Earth, Newton could resume his former ideas finding perfect agreement with experiment. He refuted the Cartesian *plenum* and the associated mechanical hypotheses, while he accepted from Descartes the principle of motion by inertia in vacuum, which Galileo had not explicitly introduced in his cosmological system. Newton placed the principle at the basis of his laws as the first and basic *lex motus*: 'Every body continues in its state of rest or uniform rectilinear motion, if not obliged to change such state by the action of imposed forces'.

The principle of inertia, fully developed in his mind since 1680, was used to explain the motion of celestial bodies in space as a consequence of inertia which makes bodies continue their rectilinear motion indefinitely, and the gravitational force between two masses, a force proportional to the masses and to the inverse of the square of their distance, that obliges every planet to deviate its trajectory along an ellipse.

The complex axiomatic construction, the ordering of the fundamental concepts of 'classical' mechanics, the development of the theorems pertinent to circular and elliptic motions and the derivation of the central forces were all completed by Newton between 1684 and 1686.

The presentation and printing of his principal work, *Philosophiae naturalis principia mathematica*, in 1687, was solicited and supported at his own expense by Edmond Halley as the Royal Society, which had promised to pay for its production, had a shortage of money. Halley (1656–1742), who later was appointed Astronomer Royal at Greenwich, is famous for his studies on comets and for having recognized that the events of the years 1456, 1531, 1607 and 1682 were due to a single comet, now known under his name, which travels in a highly eccentric ellipse with a period of about 72 years. This comet last appeared in 1985.

In the first book of *Principia*, the laws of motion, curvilinear and elliptic motions, the laws of collisions, the derivation of central forces and the motion of pendula are all contained. The second book is dedicated to the motion of solid bodies in resisting media, and implies a detailed and systematic confutation of the Cartesian physics of *plenum*, which alters the real behaviour of bodies moving inside fluids and makes indemonstrable on physical grounds the laws of Kepler. These first two books have a rational axiomatic and deductive structure; the third book starts from these premises and develops inductively the construction of the universe. The author reformulates in a simple and elegant way the Copernican heliocentric theory, adding the most recent contemporary astronomical data; after the demonstration of the laws of Kepler derived from the dynamic principles he had formulated, Newton develops the theory of the Moon's motion, of tides, the calculations relative to the trajectories of comets and the three body problem.

The first edition of *Principia* (about one thousand copies) had a large European diffusion, even if very few understood its content.

Newton as a public man

A short while after the publication of *Principia*, Newton, still short of his 50th birthday, entered politics. On the eve of the 'glorious revolution' of 1688, Newton strongly opposed the attempts made by James (II) Stuart to force the Cambridge academic corps to accept a Benedectine monk at the university with its rigid protestant traditions. The episode is at the origin of the election in 1688 of Newton as a representative of Cambridge University to Parliament, which ratified the fall of Stuart and the rise to the throne of William of Orange and composed the Bill of Rights.

After a severe neurotic crisis, in 1693, Newton was appointed warden of the London mint and directed the great operation of substituting all circulating money with new coinage. He was firmly established in the political and cultural life of the capital, revered at the Royal Court, knighted by Queen Anne, when in 1697 he left Cambridge and four years later his chair. In London he shared his Kensington quarters with his niece Catherine Barton, a brilliant and worldly lady who later married John Conduitt, the first biographer of the great scientist. Later on, in the years between 1704 and 1727, Newton was appointed president of the Royal Society.

Newton was an extremely versatile spirit who also carried out studies of theology and alchemy which he resumed in his old age and led the economist Lord Keynes in the 1930s to say: 'Newton was the last of the magicians'. In fact he fully favoured the tendency of the 17th century to combine exact sciences with magic. In his library there were 138 titles on alchemy, which was about one twelfth of all his books.

When he died on 20 March 1727, the king granted him the honour of being buried in Westminster Abbey, where he is remembered with a simple Lucretian citation: 'Newton, qui genus humanum ingenio superavit'.

His contributions to the understanding of optical phenomena were immortalized in a famous epitaph by the poet Alexander Pope (1688–1744): 'Nature and Nature's laws lay hid in night, God said, "Let Newton be" and all was light'.

Newton's theory of light

The textbook *Opticks* (1704) begins by defining the characteristics of a light ray: light rays are created in the Sun and travel to us through space; each kind of ray produces in the eye a different sensation; red, green, blue, and so on; natural light of the Sun is the sum of all these rays and appears white; the different rays can be separated by a glass prism.

Although reasonably careful to hide with skilful philosophical language his thoughts on a particular model of light, Newton was not able to resist the temptation of formulating his own point of view, by asking if light rays are made of small particles (corpuscles) emitted by the Sun and other light sources. He thought the smaller particles evoked impressions of violet and blue, and the increasingly large corpuscles induced the green, yellow, orange and red. Our sensation of colour must therefore be understood as our subjective response to the objective reality of the dimensions of the corpuscles.

In the *Principia*, Newton applies these considerations to the derivation of the Law of Refraction. Light is not only a body, but is subjected to the same mechanical laws which govern the planets' motion; if left undisturbed light propagates in a straight line, according to the law of inertia that applies to all material objects. According to Newton, light consists, therefore, of particles emitted by luminous sources, which propagate in space in a straight line. The idea recalls the Pythagorean approach, already supported by Descartes, but the points of view of the two scientists were very different.

Voltaire during his travels in England in 1728 wrote:

'A Frenchman who arrives in London, will find philosophy, like everything else, very much changed there. He had left the world a plenum, and he now finds it a vacuum. At Paris the universe is seen composed of vortices of subtle matter; but nothing like it is seen in London. In France, it is the pressure of the moon that causes the tides; but in England it is the sea that gravitates towards the moon. [...] You will observe further, that the sun, which in France is said to have nothing to do in the affair, comes in here for very near a quarter of its assistance. According to your Cartesians, everything is performed by an impulsion, of which we have very little notion; and according to Sir Isaac Newton, it is by an attraction, the cause of which is as much unknown to us. [...] A Cartesian declares that light exists in the air; but a Newtonian asserts that it comes from the sun in six minutes and a half.'

And added:

'Very few people in England read Descartes, whose works indeed are now useless. On the other side, but a small number peruse those of Sir Isaac, because to do this the student must be deeply skilled in the mathematics, otherwise those works will be unintelligible to him. But notwithstanding this, these great men are the subject of everyone's discourse. Sir Isaac Newton is allowed every advantage, whilst Descartes is not indulged a single one. According to some, if one has no more horror of the vacuum, if one knows that air is weighted, if one uses glasses, all this we owe to Newton. Sir Isaac Newton is here as the Hercules of fabulous story, to whom the ignorant ascribed all the feats of ancient heroes.'

Some one hundred years later, on 28 December 1817, a group of poets met in London in the studio of the painter Benjamin Haydon, including Words-worth, who abused Haydon for painting Newton's head in his picture, saying that he was 'a fellow who believed nothing unless it was as clear as the three sides of a triangle' and Keats, who added that Newton had destroyed all the poetry of the rainbow by reducing it to its prismatic colours. They all toasted 'Newton's health and confusion to mathematics'.

The wave theory becomes predominant *pro tempore*
Both wave and corpuscular theories, although seemingly incompatible, gave rise to a series of heated scientific quarrels between their respective supporters, until the experiments and the theoretical considerations of T Young (1773–1829), E L Malus (1775–1812), L Euler (1707–1783), A Fresnel (1788–1827), J Fraunhofer (1787–1826) and others, confirmed the first theory.

Leonard Euler—the great Swiss mathematician, a member of the Science Academies of Poland and Russia—captured the attention of Europe with a series of letters, published in 36 editions and nine languages, written between 1760 and 1762 to the German princesses of Anhalt-Dessau who asked him his opinion on every aspect of science. Writing about sunlight he asks the question: 'What are these rays? This is with no doubt, one of the most important questions in physics' and added, so supporting without discussion the wave theory, that the Sun's rays are 'with respect to the aether, what sound is with respect to the air'.

Maxwell's theory of electromagnetism
One hundred years later, in 1864, J C Maxwell (1831–1879) discovered the electromagnetic and inelastic nature of light vibrations, summarizing it in

the famous equations which bear his name and describe the different electric and magnetic phenomena (electromagnetism) in a general form which allow the existence of light to be predicted. The electromagnetic waves are made by vibrations in time and space of both electric and magnetic fields, and travel at the impressive speed of $300\,000\ \mathrm{km\ s^{-1}}$, the same velocity with which, according to previous measurements done already in 1675 by Roemer, and later on with greater precision by H L Fizeau (1819–1896) in 1849, light travels. With his equations, Maxwell also suggested the way to produce artificially these waves, and in fact in 1887, H Hertz (1857–1894) succeeded in producing electromagnetic waves of wavelength of a few metres.

James Clerk Maxwell is considered together with Newton and Einstein to be one of the three great geniuses of physics. Not by chance did Einstein have a portrait of Maxwell in his study in Princeton. Maxwell was born in Edinburgh, in Scotland, in a middle class family. His father, John Clerk, was a lawyer who inherited the Maxwell estate in Scotland, together with their family name, and built a house in the locality of Glenlair where the family moved a little after the birth of James. When he was eight, his mother (to whom he was very close) died, and he remained with his beloved father who never remarried. Maxwell was able to draw, wrote verses and loved animals. Being of delicate health, he was often sick. He was interested in mathematics and geometry when still a schoolboy, and a teacher at Edinburgh University, Professor James D Forbes (1809–1868), who was his intellectual guide for many years, presented one of his first mathematical works on the description of some curves to the Edinburgh Royal Society, as early as 1846. He went to the local University from 1847 to 1850. In 1849, his mathematics professor Kelland presented to the Edinburgh Royal Society another work on curves and, in 1850, one more on the equilibrium of elastic solids. During those years, Maxwell was also interested in colour vision. He went to Cambridge in 1850 and was appointed fellow in 1855. Here he was a member of the exclusive Apostle Club, giving several speeches which testify to his many interests of ethic-philosophic, religious, logic and methodological character. While he was studying for the examination to become fellow, he became interested in electricity and magnetism. At that time, by studying fish eyes, he succeeded in describing mathematically the optical properties that a medium must have to obtain perfect focusing. About 50 years later, R K Luneburg rediscovered the subject by considering a lens with the properties studied by Maxwell. After Maxwell obtained the nomination to become a fellow, no position being available at Cambridge, he went to Scotland where from 1856 to 1860 he was professor of natural philosophy at Marischal College of Aberdeen, having chosen this place because it gave him the chance to enjoy a long summer vacation which allowed him to pass six months per year in his Glenlair estate. During that period he married. One of his students in Aberdeen, David Gill (1843–1914), afterwards a pioneer of astronomical photography and Astronomer

Royal, describes Maxwell's lessons as follows:

> 'In those days a professor was little better than a schoolmaster—and
> Maxwell was not a good schoolmaster; only some four or five of us, in
> a class of seventy or eighty, got much out of him. We used to remain
> with him for a couple of hours after lectures, till his terrible wife came
> and dragged him away to a miserable dinner at three o'clock in the
> afternoon. By himself he was the most delightful and sympathetic of
> beings—often dreaming and suddenly awakening—then he spoke of
> what he had been thinking. Much of it we could not understand at the
> time, and some of it we afterwards remembered and understood'.

Research on the stability of the rings of Saturn allowed him to win in 1857 a
prize and revealed him as one of the best mathematical physicists of his time.
From the analysis of the problem, he derived that the rings are formed by
many particles, as we know today from better astronomical observations.
From 1860 to 1865 he was at King's College in London where he elaborated
on his principal works and met and frequented Michael Faraday (1791–
1867), the father of electrology, for whom he had a great admiration and
considered to be most learned in the subject of electricity and magnetism.

In 1865 he became sick and retired for six years to his house in Glenlair,
coming out only for short journeys, one of which was to Italy in 1867. At
Glenlair he finished his kinetic theory of gases and wrote his famous textbook
Treatise on Electricity and Magnetism which contains, fully developed, the
electromagnetic field theory. With respect to the equations which synthesize
all his work, Einstein said: 'Special relativity owes its origins to Maxwell's
equations of the electromagnetic field' and Boltzmann himself asked 'Is it
perhaps a God who wrote these symbols?'

In 1871 he accepted the chair in experimental physics just established in
Cambridge, and the directorship of the Cavendish Laboratory. The laboratory
was already established when in October 1870 the Duke of Devonshire, Dean
of Cambridge University, expressed the desire to support the construction of a
physics laboratory and the provision of its equipment. The laboratory was
named after one of the Duke's relatives, Henry Cavendish (1731–1810) who
had dedicated all his life to chemistry and experimental physics, being interested mostly in electricity. Maxwell worked actively towards the organization
of the laboratory yet also found the time for his textbook which he published
two years later (1873). He died in Cambridge on 5 November 1879.

To him we owe fundamental contributions to the theory of electromagnetism, thermodynamics and the kinetic theory of gases of which he
may be considered one of the founders together with Ludwig Boltzmann
(1844–1906) and Josiah Willard Gibbs (1839–1903). The kinetic theory
considers a gas to be composed of a great number of atoms or molecules
which move freely in the available space, colliding with each other and
with the walls of the container. Through this model the theory allows us to
interpret the macroscopic properties of gases. John Herapath (1790–1868)

was the first to establish a link between the gas temperature and the velocity of its molecules, even if the relation he found was wrong. The subject was studied by the Englishman James P Joule (1818–1889) and by the Germans Rudolf Clausius (1822–1888) and Ludwig Boltzmann. Maxwell derived from theory certain properties of gases, yielding the molecular velocity distribution law, by assuming the molecules behave as small billiard balls. He obtained expressions for the pressure, viscosity, diffusion etc. He derived the energy equipartition theorem of which we will speak later. Maxwell considered two possible methods to describe a gas. One is based on the laws of dynamics and describes the deterministic behaviour of the single constituents of the gas giving the full description of the system. The other method is statistical in nature and discounts knowledge of the behaviour of single molecules, given their enormous number in a gas; it treats the system by utilizing the laws of statistics which allow quantities to be derived that describe globally the behaviour of the gas such as pressure, temperature etc.

Maxwell was also interested in the theory of colours, and developed and completed the theory of the physician and physicist Thomas Young, who claimed that coloured vision can be obtained by combining three images in the fundamental colours to which, in the human eye, correspond three different kinds of receptors. Maxwell identified the three primary colours from which all other colours can be obtained as red, blue and green, and indicated, as a cause of colour blindness, the absence in the affected eye of one of the three kinds of receptor. He concluded that making a photograph through filters of those colours and then recombining the images, a coloured photograph of any subject should be obtained. He gave a practical demonstration during a conference at the Royal Society in 1861 by making a photograph of a Scottish ribbon with a method that is substantially the same as that employed today. It was the first colour photograph.

The electromagnetic field theory is, however, the most important result obtained by Maxwell and without doubt one of the most significant advances in science on which the science and technology of today are based.

Electromagnetism, in the middle of the 19th century, consisted of a great quantity of experimental results to which Faraday had contributed greatly, but was still waiting for a general theory capable of their interpretation.

Michael Faraday (1791–1867) is an extraordinary example of an innovative researcher. He was the son of a blacksmith and started to work at 13 as an apprentice near a bookbinding shop. Here he started to read works of chemistry and electricity and perform experiments with self-made equipment. In 1813, he met the chemist Humphrey Davy (1778–1829) and became his assistant at the Royal Institution. He was a very skilled experimentalist and discovered fundamental phenomena that provided the basis for the electromagnetic theory of Maxwell. He developed a method to visualise the force lines of electric and magnetic fields. As Davy's assistant he travelled frequently (1813–1815) in Europe which exposed him to the works of the most important researchers of the continent.

In 1821, extending the experiments of the Dutch physicist H C Oersted (1777–1851) he showed that magnets exert mechanical actions on conductors in which an electric current flows. After this, he studied electrolysis phenomena, enouncing the laws which they obey. In 1830–31 he discovered the phenomenon of electromagnetic induction. The action exerted by a magnetic field on polarized light (Faraday effect) and diamagnetism were among his last discoveries. In 1862 he tried to study the effects of a magnetic field on the spectra of light, pioneering research which was later completed successfully by P Zeeman.

Maxwell interpreted in a brilliant way the results of Faraday and other researchers, showing that electric and magnetic phenomena were intimately connected and that in some cases an electromagnetic field could propagate as a wave, realizing that light is a wave of this kind.

Maxwell's electromagnetic theory was accepted with great reluctance. Even Maxwell and his pupils tried for a long time to describe the electromagnetic field by means of mechanical models. Only gradually was an explanation of his equations in terms of mechanical models given up and eventually the concept that electric and magnetic fields are real quantities was accepted.

The wave theory considers light as vibrations of the aether, explaining wonderfully well reflection, refraction, diffraction and interference and other phenomena, and the property of light to appear coloured is put in context with the properties of the wave, white light being nothing more than the superposition of all colours (a fact for which Newton had already given an experimental demonstration). A given colour is defined by the wavelength of the radiation (figure 6), that is the distance between two successive peaks of the wave. This wavelength is measured, in the visible region, in

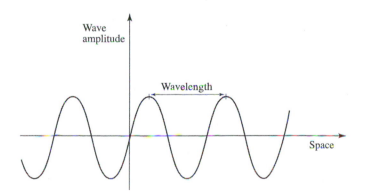

Figure 6. A sinusoidal wave seen at a given time as a function of position.

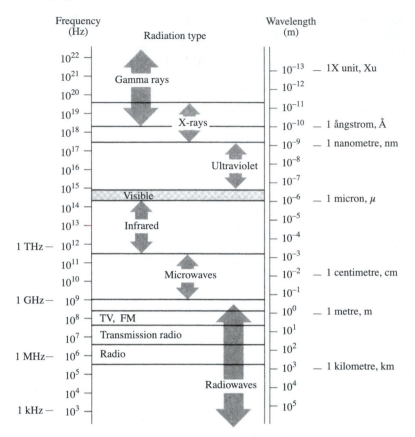

Figure 7. The electromagnetic spectrum. The labelling on the left gives the frequency. On the right, the corresponding wavelengths are shown.

angstrom[1] (one ångström or Å is a hundredth millionth of a centimetre) so that the visible region extends from about 3800 Å (violet) to 7000 Å (red). The number of peaks of the wave which pass through a point in a second is the frequency of the wave and is measured in hertz (Hz). The product of the wavelength and the frequency of a wave is equal to its propagation velocity. So, for example, green light which in vacuum has a wavelength of 5500 Å and propagates at the speed of 300 000 km s^{-1}, has a frequency of 545 000 billion hertz. Radiations of greater wavelength follow the sequence infrared, microwaves and radio waves. Those of shorter wavelength are ultraviolet, x-rays and gamma rays (figure 7).

---

[1] According to the International System of Units one should use the nanometer (nm), that is the thousandth millionth part of a metre, being 1 nm = 10 Å, but in that time ångströms were used.

CHAPTER 2

SPECTROSCOPY

If the light emitted by the Sun or by an incandescent lamp is incident onto a glass prism, we see— just as Newton did—the colours displayed in a sequence from violet to red; Newton called this the 'spectrum'. The word has remained to indicate the image obtained by decomposing any light with a prism or any other more complex apparatus. When the intensity varies gradually from one colour to another, we speak of a 'continuous spectrum'. Generally, light produced by an electric discharge in a gas (as in neon light signs) exhibits very bright lines on a black background; in this case the spectrum is referred to as a 'line spectrum'.

In spectroscopes, the instruments used today to study such spectra, light passes through a thin slit positioned at the entrance of the instrument, and in the plane of observation every line is the image of the entrance slit relative to that monochromatic (i.e. single colour) component of the radiation under study. Each one of these lines has a well-defined position and intensity which establish the characteristics of the spectrum. The position is determined by the wavelength, and therefore by the frequency of the monochromatic radiation which has produced the line.

If we consider substances made by free atoms of the same species, that is if we consider elements in the gaseous state, we find that their spectra are essentially line spectra. More exactly, these spectra contain lines of gradually decreasing wavelength that follow one another and become more and more finely spaced until below some wavelength they merge into a continuous spectrum.

The dawn of spectroscopy
The study of the composition of light emitted by incandescent bodies is the subject of spectroscopy. This discipline was born in the 19th century and has played a fundamental role in the study of light emission and of the structure of atoms, and is an indispensable tool for understanding the working principles of masers or lasers. We may say that it started in 1802 with the discovery, made by the English physician William Hyde Wollaston (1766–1828), of the presence of dark lines in the spectrum of the light from the Sun.

Wollaston became a rich man by inventing in 1804 a process to produce pure, malleable platinum suitable for making vessels; he also isolated two new elements, palladium (in 1804) and rhodium (in 1805), the former named in honour of the asteroid Pallas, also discovered in 1804, and the latter for the pink colour of its compounds. In 1807 Wollaston patented the lucid camera, in which a prism reflects light coming from the object to be drawn onto paper and thus into the eye of the artist. The artist therefore attains the illusion that the image is already on the paper and may draw it very simply following its contours. Wollaston was a friend of Thomas Young and was a supporter of the wave theory of light. When in 1802 he observed the dark lines in the spectrum of the Sun, he did not understand their importance, believing they were simply the natural contours of the line colours.

Twelve years later Joseph Fraunhofer (1787–1826), an employee of the Benediktbeuern Opto-Mechanics Institute in Bavaria, while measuring the dispersive power of several different glasses, rediscovered the dark lines in the solar spectrum and started to study them. He found the position in the spectrum of a great number of them (576, to be exact), and denoted the most prominent ones by letters from A in the extreme red to H in the violet (figure 8). The dark lines indicate that certain wavelengths are absent in sunlight when it reaches the earth. At the same time Fraunhofer discovered that the bright yellow line (actually two lines very near to each other), which is present in the light of all flames observed with a spectroscope, is in the same position of the dark line that he identified with the letter D in the solar spectrum.

Fraunhofer was the son of a poor glass cutter. First he was an apprentice in a grinder's shop and then a worker in a mirror factory. Later, the factory collapsed and the 15 year old boy lay buried under the debris. He was taken out half dead and was given by the king, as a gift for his miraculous escape, 18 ducats: a real treasure for him. With the money he bought instruments and books and later became a famous optician.

He had the great ambition to improve the achromatic lenses (lenses free from dispersion effects) that the English optician John Dollond (1706–1761) had studied. In 1756, Dollond put together two glasses with almost opposite dispersive powers in order to eliminate chromatic aberrations. The greatest difficulty was to measure the dispersive power of different kinds of glasses

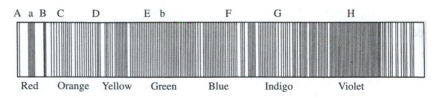

Figure 8. Solar spectrum with Fraunhofer dark lines, some of which are shown with the lettering used by Fraunhofer.

used in these lenses and, by measuring various glass prisms, Fraunhofer discovered the dark lines in the solar spectrum.

His discovery represented the forerunner of what later was called 'spectral chemical analysis', which may be considered to have started in 1826, when William Henry Fox Talbot (1800–1877) found that a precise relation exists between the luminous spectrum of a flame and the substances contained within it, suggesting that the colour of the light emission could be used, instead of a long chemical analysis, to ascertain the nature of the burning substances.

At the beginning of the 19th century the instruments (spectroscopes) needed to measure the structure of the spectra with sufficient precision, and the methods to measure the wavelengths of different radiations were sufficiently well developed, mostly thanks to the work by J Fraunhofer and A Fresnel, and therefore Talbot's suggestion could be put into practice.

Figure 9 shows the simplest type of spectroscope, like those still used in schools. The core of the instrument is a glass prism lying between two small telescopes. One is fitted with an adjustable thin slit and allows the light under examination (in the figure a flame) to fall on the prism; the second telescope collects the dispersed light. A third telescope is actually a simple tube which bears at its end a small scale with cross-hairs illuminated from the outside, and whose image magnified by a lens is projected onto the spectrum to give a reference for the position of the different lines.

The final conclusive steps towards spectral chemical analysis are credited to the chemist Robert Bunsen (1811–1899) and the physicist Gustav Kirchhoff (1824–1887), who in 1860–61 were colleagues at Heidelberg University. They

Figure 9. Old model of a spectroscope with Bunsen's burner. (From B H Burgel 1946 *Dai mondi lontani*, Einaudi, Torino, p 113).

33

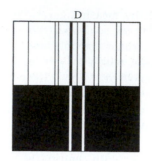

Figure 10. Line inversion. On the upper part the sodium D (doublet) lines appear black in the solar spectrum. Below, the D lines of the sodium vapour in a laboratory flame appear bright.

built a standard instrument to analyse the spectra of the elements contained in salts which were made incandescent by a flame (the so-called Bunsen burner which was essentially a gas flame in which the bodies under examination were burnt) and with it discovered that the luminous lines emitted in the spectrum of every examined metal were characteristic of the metal itself. Their discovery of the new elements caesium and rubidium was the first practical proof of the enormous possibilities of spectroscopic analysis for establishing the chemical composition of matter. By using this technique, Kirchhoff identified many of Fraunhofer's dark lines. The D-line, for example, is due to sodium (figure 10). David Brewster (1781–1868) in 1832 had already explained the origin of the dark lines in the solar spectrum by noticing that the light emitted by the hot surface of the Sun, before reaching Earth, traverses the outer cooler atmosphere of the Sun whose components absorb practically the same wavelengths that they emit when excited. Therefore a dark line is observed where light would be expected if it were not for the solar atmosphere. Taking this further, sodium vapour for example (easily obtained by putting some sodium chloride, i.e. normal kitchen salt, in the flame of a Bunsen burner) emits a characteristic yellow light, produced by two narrow lines very close to each other (the D-line). Whenever these lines are found in a spectrum, we may confidently infer that sodium is contained in the source. In this way a new extremely powerful method of chemical analysis was born, which allows the detection of minimal traces of a substance, and being independent from the source distance, may be applied to the study of very far bodies such as stars.

The wavelengths of radiation that a substance emits are also the ones it may absorb. If, for example, we send an intense light containing all visible wavelengths (a continuum) through a flame in which sodium is burnt, the transmitted light is found to lack the wavelengths corresponding to the two yellow lines of the D-line of sodium. In the spectrum two dark lines appear at the place of the two bright lines that are observed in the emission spectrum. This explanation applies not only to our Sun but to every star.

In fact, dark lines similar to the ones Fraunhofer observed in the solar spectrum are observed in the spectra of all stars, and their positions in the spectrum indicate which wavelengths have been absorbed by the substances present in the stellar atmospheres, so allowing us to identify them.

Fundamental contributions to this new science of spectroscopy were made by Kirchhoff, David Brewster, John Herschel (1792–1871), William Henry Fox Talbot, Charles Wheatstone (1802–1875), Antoine-Philibert Masson (1806–1880), Anders Jonas Ångström (1814–1874) and William Swan.

The fact that spectra of a substance are sometimes composed of a number of discrete lines and sometimes occur in the form of bands was finally explained by George Salet (1875)—after many controversies—by associating the line and band spectra to atoms and molecules respectively.

Atoms

Already Democritus and Leucippus in the 5th century BC had spoken of atoms. The Latin poet Lucretius (98–55 BC) in *De rerum natura*, explaining Democritus' theory, said that air, water, earth and all the other things of this world are made by a number of particles or corpuscles—atoms—engaged in a restless, very rapid movement, and these are so small as to be perfectly invisible to the human eye. Atoms were assumed to be the ultimate result of the subdivision of matter, the word 'atom' in Greek meaning 'indivisible'.

However, the ideas of both Democritus and Lucretius were far removed from our present way of considering atoms, because they did not deduce that there existed so many different kinds of atoms, and that atoms of a given species were all identical.

Expressed in a vague and generic way, Democritus' atomic theory languished and the word 'atom' became used only to represent an object of extremely small dimensions. Lucretius achieved success in a very different field by hypothesizing that infective diseases were diffused by very small particles, and in the Renaissance the Italian philosopher and physician Girolamo Fracastoro (1483–1553) resumed his theory. Afterwards, however, the secret of bacteriology was completely forgotten until, eventually, it was resurrected by Louis Pasteur (1822–1895).

In any case, the atomic hypothesis of Democritus and Leucippus was adopted by the priest Gassendi (1592–1655) who worked hard to demonstrate that, even if it was promulgated by the 'dissolute' Epicurus and Lucretius, it had no direct connections with the religious philosophies of its ancient supporters, and could be accepted even by a Christian man. Gassendi's ideas had a profound influence on the chemist Robert Boyle to whom we owe the notion of the chemical element, a notion which became more precise when Antoine-Laurent Lavoisier (1743–1794) discovered that

35

chemical compounds with a definite individuality contain chemical elements in constant proportions, and that the total mass in a chemical reaction is the same before and after the reaction. This research culminated in the work of John Dalton (1766–1844) who exposed with great clarity in his book *A New System of Chemical Philosophy* published in 1808 in Manchester, the concept that there are many kinds of atoms, each one characterizing a different substance, and that the atoms of a given species are all identical. Dalton proved that to each chemical element could be assigned a number which represented the weight of one atom or of a molecule of the element.

The atomic weight

Today the atomic or molecular weight is measured by comparing the weight of an atom of the substance we are interested in with the weight of a hydrogen atom. The hydrogen atom weight is conventionally established to be approximately one atomic mass unit (one atomic mass weight is 1.66×10^{-24} g), in such way that the weight of the carbon atom is just 12. The quantity of a substance whose weight in grams is equal to its atomic or molecular weight is called a mole. Therefore a mole of carbon atoms weighs 12 g. The atomic weight of oxygen is nearly 16 (15.9994 exactly) and because a molecule of oxygen contains two atoms, a mole of oxygen molecules weighs nearly 32 g. From this definition of a mole it follows that the mole contains always the same number of atoms or molecules (the so-called Avogadro's number = 6.022×10^{22}).

However, the existence of atoms was still controversial because nobody could say he had seen one, and many different ideas about them were circulating. In 1860 at the Chemistry Congress in Karlsruhe, the first international conference ever held, participants discussed at length the difference between atoms and molecules without reaching any definite conclusion. Today we may define a molecule as a chemical combination of two or more identical or different atoms that may exist as a stable entity and which represents the smallest quantity of matter that holds the characteristic properties of the substance in question.

The electron shows up

While the atomic and molecular theories of chemistry were growing, studies on electrical conduction in liquids and on the properties of electrical discharges in gases at low pressure revealed that really the atom is not 'indivisible', but contains in its interior electric charges. G J Stoney (1826–1911) in 1874, in attempting to explain in a simple way the laws of electrolysis put forward by M Faraday in 1833, introduced the 'atom' of electricity which later (1891) he named the 'electron'. Previously, H L F Helmholtz (1821–1894) had said 'If we accept the hypothesis that the elementary substances are composed of atoms, we cannot avoid concluding that electricity also, positive as well as negative, is divided into definite elementary portions which behave like atoms of electricity'.

To the so-conceived electron only the property of representing the electrical charge was granted. However, the English physicist J J Thomson (1856–1940) discovered in 1897 that in the electrical discharges produced in gases very small particles with negative charge were present, which were emitted from atoms. Thomson succeeded in measuring their specific charge (e/m), that is the ratio of their charge to their mass. Because the atom by itself is neutral, the discovery of the electron—for which Thomson in 1906 was awarded the Nobel prize for physics –inferred that the atom should contain in its interior both negative (electrons) and positive charges.

The Nobel prize

As many of our characters have been awarded this esteemed prize, it is worth making a small digression and describing how it was born and how it is assigned.

On 10 December 1896, the Swedish chemist Alfred Nobel died of a stroke at 63 in his house in San Remo. He had created practically-useful nitroglycerin, discovered by A Sobrero (1812–1888), stabilizing it with an inert absorbing substance, and in 1875 had produced the first dynamite by gelatinizing colloidal cotton with nitroglycerin. In 1889, mixing different proportions of colloidal cotton with nitroglycerin, he was able to fabricate in Avigliana, near Turin in Italy, the ballistite that is still one of the most commonly employed non-deflagrating explosives. Thanks to his patents and also to the exploitation of oil fields acquired in Baku, Nobel realized an immense fortune of several hundreds of millions of euros, which—as if trying to make up for the aggressive use mankind was to make of his inventions—he bequeathed to a foundation. The pertinent excerpt from his will recites textually:

> 'The whole of my remaining realizable estate shall be dealt with in the following way: the capital, invested in safe securities by my executors, shall constitute a fund, the interest of which shall be annually distributed in the form of prizes to those who, during the preceding year, shall have conferred the greatest benefit on mankind. The said interest shall be divided into five equal parts, which shall be apportioned as follows: one part to the person who shall have made the most important discovery or invention within the field of physics; one part to the person who shall have made the most important chemical discovery or improvement; one part to the person who shall have made the most important discovery within the domain of physiology or medicine; one part to the person who shall have produced in the field of literature the most outstanding work in an ideal direction; and one part to the person who shall have done the most or the best work for

fraternity between nations, for the abolition or reduction of standing armies and for the holding and promotion of peace congresses. The prizes for physics and chemistry shall be awarded by the Swedish Academy of Sciences; that for physiology or medical works by the Karolinska Institute in Stockholm; that for literature by the Academy in Stockholm, and that for champions of peace by a committee of five persons to be elected by the Norvegian Storting. It is my express wish that in awarding the prizes no consideration be given to the nationality of the candidates, but that the most worthy shall receive the prize, whether he be Scandinavian or not.'

Each prize was worth the present-day equivalent of about 50 000 euros, and surely it was the richest at the time.

The rules for the selection of the winners are similar for the different fields. In the case of physics, a scholarly group made of the former winners of the prize and other well known people, members of the Swedish Academy of Sciences, professors in Swedish universities, in the ones of Copenhagen, Helsinski, Oslo, and in six more universities that change every year, makes proposals. The proposals are evaluated by a five-member committee formed by the president of the Nobel Institute for physics and four members elected by the physical section of the Academy. The proposals of the committee are eventually examined in a meeting of the Academy. The members of the Academy receive a gold medal every time they participate in a meeting to vote, and the commission members are generously rewarded. No minutes of the meetings are made and members are forbidden to reveal details of the discussions. The final decision has to be taken before 15 November. Prizes are awarded in Stockholm and Oslo, for the peace prize, with an official ceremony on 10 December, the anniversary of the death of the founder. The winner gives a public lecture and is presented with the award by the King of Sweden.

According to the will of Alfred Nobel the prize must be assigned to the person who with his or her discovery or invention has given the greatest benefit to mankind. At the beginning this created notable difficulties in assigning the prize to theoretical researchers. The first winner for physics was, in 1901, Wilhelm Conrad Roentgen (1845–1923) 'in recognition of the extraordinary services he has rendered by the discovery of the remarkable rays subsequently named after him'.

The Balmer formula

Salet's hypothesis that spectra were characteristic of atoms or molecules was fully justified by the electromagnetic field theory then developed by Maxwell, according to which oscillating electric charges emit radiation. A search began to discover if a relationship existed among the spectral lines of the same substance or between lines of different substances.

On 25 June 1884 the Swiss Johann Jakob Balmer (1825–1898) presented a paper on this subject to the Naturforschende Gesellschaft of Basel. Balmer, a 60 year old professor in a girl's school in Basel, was strongly interested in numbers. He had reconstructed the Temple of Jerusalem from the measurements of the Ezekiel's vision; he had been interested in the number of steps of the great pyramid, and so on. Once, chatting with a friend, he complained of having nothing to do, and the friend replied: 'Well, being so interested in numbers, why do you not see what can you do with the numbers that come out of the spectrum of the simplest element, hydrogen?' His friend gave him the wavelengths of the three lines known at that time in the visible region of the spectrum. By studying the problem, Balmer realized that it was possible to represent analytically the succession of the three lines by the relation

$$\lambda = 3.36 \times 10^{-5} \frac{n^2}{n^2 - 4}$$

where λ is the wavelength expressed in centimetres and n is an integer number that may be 3, 4 or 5. Alternatively the formula may be written by using the wavenumber ν_0, that is the inverse of the wavelength and represents the number of full waves that can be counted in a centimetre. This wavenumber multiplied by the velocity of light, indicated with the letter c, gives the frequency f, or the number of vibrations per second. By using the wavenumber the Balmer formula can be written as

$$\nu_0 = 1/\lambda = 109\,678(1/4 - 1/n^2)$$

where λ is still the wavelength expressed in centimetres and the number 109 678 was later indicated with the letter R, and received the name 'Rydberg constant'. The formula not only gave the wavenumber or the wavelength of the three lines known at the time, but as new lines were discovered, gave also their values (figure 11).

During the eclipse of the Sun in 1898, 29 lines of the Balmer series were measured: all the ones that are obtained by the formula for n running from 3 to 31.

Rydberg and the combination principle

In 1886, Alexander S Herschel, the son of the great astronomer John, and Henri Deslandres (1853–1948) found a mathematical description for various band spectra. Moreover, the Swedish mathematical-physicist Johannes Robert Rydberg (1854–1919) published the results of an analysis of spectra he performed (1890) which showed that the Balmer spectral series in the visible, and other line series of hydrogen in the ultraviolet and infrared, could be represented via a general relation, which today is known by his name.

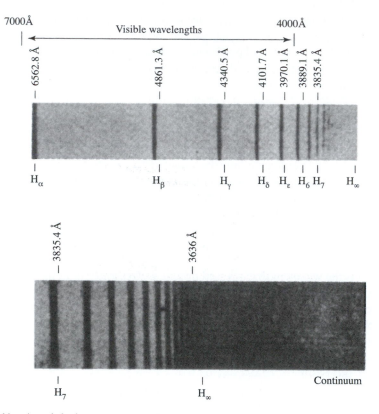

Figure 11. Atomic hydrogen spectrum. On the upper side the spectrum shows the Balmer lines in the visible; in the lower side an enlargement of the final part on the right of the spectrum shows the Balmer lines in the ultraviolet.

Rydberg was interested in the debate concerning spectra and started his studies on the relations existing among the spectral lines long before Balmer published his formula. He was interested in the periodic classification of the elements, introduced by the Russian chemist Dmitrij Mendeleev (1834–1907). Mendeleev had observed that the only method to classify chemical elements was to consider their atomic weights and also that when elements are put in order of increasing atomic weight, they show a marked periodicity in their properties, so that by ordering them in rows of increasing atomic weight, columns of chemical elements with similar properties are obtained (table 1). The great intuition of Rydberg was to grasp that the periodicity was a result of the atomic structure.

In the period between 1882 and 1887, when he was an assistant at the Physics Institute of the University of Lund, he studied the dependence of the physical and chemical properties of elements on their atomic weight, by considering the atomic weight as the principal parameter from which these

Table 1. Mendeleev's periodic table of elements.

IA	IIA	IIIA	IVA	VA	VIA	VIIA	VIIIA			IB	IIB	IIIB	IVB	VB	VIB	VIIB	
H 1 1.0079																	**He 2** 4.0026
Li 3 6.941	**Be 4** 9.01218											**B 5** 10.81	**C 6** 12.011	**N 7** 14.0067	**O 8** 15.9994	**F 9** 18.9984	**Ne 10** 20.179
Na 11 22.989	**Mg 12** 24.305											**Al 13** 26.9815	**Si 14** 28.0855	**P 15** 30.9738	**S 16** 32.06	**Cl 17** 35.453	**Ar 18** 39.948
K 19 39.098	**Ca 20** 40.08	**Sc 21** 44.956	**Ti 22** 47.90	**V 23** 50.9415	**Cr 24** 51.996	**Mn 25** 54.938	**Fe 26** 55.847	**Co 27** 58.9332	**Ni 28** 58.70	**Cu 29** 63.546	**Zn 30** 65.38	**Ga 31** 69.72	**Ge 32** 72.59	**As 33** 74.9216	**Se 34** 78.96	**Br 35** 79.904	**Kr 36** 83.80
Rb 37 85.468	**Sr 38** 87.62	**Y 39** 88.906	**Zr 40** 91.22	**Nb 41** 92.9064	**Mo 42** 95.94	**Tc 43** 98	**Ru 44** 101.07	**Rh 45** 102.906	**Pd 46** 106.4	**Ag 47** 107.868	**Cd 48** 112.41	**In 49** 114.82	**Sn 50** 118.69	**Sb 51** 121.75	**Te 52** 127.60	**I 53** 126.905	**Xe 54** 131.30
Cs 55 132.905	**Ba 56** 137.33	**La 57** 138.906	**Hf 72** 178.49	**Ta 73** 180.948	**W 74** 183.85	**Re 75** 186.207	**Os 76** 190.2	**Ir 77** 192.22	**Pt 78** 195.09	**Au 79** 196.967	**Hg 80** 200.59	**Tl 81** 204.37	**Pb 82** 207.2	**Bi 83** 208.98	**Po 84** 209	**At 85** 210	**Rn 86** 222
Fr 87 223	**Ra 88** 226.025	**Ac 89** 227.028															

Ce 58 140.12	**Pr 59** 140.908	**Nd 60** 144.24	**Pm 61** 145	**Sm 62** 150.4	**Eu 63** 151.96	**Gd 64** 157.25	**Tb 65** 158.925	**Dy 66** 162.50	**Ho 67** 164.93	**Er 68** 167.26	**Tm 69** 168.934	**Yb 70** 173.04	**Lu 71** 174.967
Th 90 232.038	**Pa 91** 231.036	**U 92** 238.029	**Np 93** 237.048	**Pu 94** 244	**Am 95** 243	**Cm 96** 247	**Bk 97** 247	**Cf 98** 251	**Es 99** 252	**Fm 100** 257	**Md 101** 258	**No 102** 259	**Lr 103** 260

⟵ Symbol and atomic number
⟵ Atomic weight

Solid | *Liquido* | *Gas*

41

properties depend, and started to study the relationships existing among the spectral lines of elements. The problem he wanted to solve required a systematic search through existing spectroscopic material in order to establish a semi-empirical formula that modelled all the data universally.

The history of science shows that any field of physics passes through a phase in which accumulated empirical material is subjected to a 'preliminary processing' activity during which some general laws are extracted, even if the phenomena under study do not yet have a theoretical base. Such examples are Kepler's laws in celestial mechanics or the Boyle–Mariotte law for gases.

These 'primary' theoretical models fulfil a double function: first, they create a certain general basis for systematizing experimental data, playing the role of empirical laws. Then they have an important active role in creating more fundamental theories, performing the role of constructive mediators between theoretical knowledge and empiricism. Thus, for example, Maxwell in the construction process of the theory of electromagnetism did not consider directly experimental data, but used theoretical knowledge of the preceding level (Biot-Savart's law that yielded the magnetic field created by a wire in which a current flows, Faraday's law of induction etc.) as selected 'empirical' material.

If we consider from this point of view the status reached by spectroscopy in the 1880s, one must recognize that to find the laws governing the spectral lines was the most important problem at that time. In such cases, a situation often results that is frequent in the development of science and that we will encounter at other times. Different researchers attempt independently to find a solution to the same problem and find its solution simultaneously. This circumstance happened in this case. Independently from Rydberg, in 1890, two well known spectroscopists, Heinrich Kayser (1853–1940) and Carl Runge (1856–1927) tried to establish some general mathematical equations concerning the laws that govern spectroscopy and suggested solutions which were discussed with bright polemics, until the point of view of Rydberg prevailed, reaching the greatest recognition by the end of the century.

According to Rydberg, the analytical expression for the spectra had to be a function of an integer number. He was studying which form this function should have, and found one in which the inverse of the wavelength depended on the inverse of an integer number squared, when Balmer published his formula for the hydrogen atom, which corresponded to a particular case of the Rydberg expression.

Kayser and Runge, on the contrary, were looking for an algebraic expression which would predict with the greatest precision for the inverse of the wavelengths of the series, and found one which used the inverse of an integer number squared and the inverse of an integer to the fourth power. Although they admitted that Rydberg was right in saying that their expression was simply one among many which could be written, they objected that among all of them it was the most accurate. The fact that

Rydberg maintained that the relation to be found should have a universal validity, for all atoms, did not interest them in any way.

The Rydberg representation gave the inverse of the wavelength of the spectral lines of atomic spectra within the same series, as the difference between two 'spectral terms' (as they were later called) which were each in the form of a universal constant (later called in his honour the 'Rydberg constant') divided by the square of the sum of an integer number plus a constant, typical of the series. In this formulation the 'combination principle' later expressed by the Swiss scientist Walther Ritz (1878–1909) was already present.

At the time, the luminous vibrations represented by line spectra were assumed to be produced all together in the atom. Eventually, in 1907, Arthur William Conway (1875–1950) a mathematical-physics professor in Dublin, gave the right explanation according to which the atom produces spectral lines one at a time, so that the production of the complete spectrum depends on the presence of a large number of atoms. According to Conway, to generate a spectral line the atom should be in an abnormal or disturbed state. This situation, in which a single electron situated within the atom is stimulated to produce vibrations of the frequency corresponding to the spectral line in question, does not endure indefinitely, but lasts only a time sufficient to enable the electron to emit a train of vibrations.

These ideas were reaffirmed in 1910 by Penry Vaughan Bevan (1875–1913) who also concluded that spectroscopic phenomena must be explained by the presence of a large number of atoms which at any one instant are in different states and each of which at that instant is concerned not with the whole spectrum but at most with only one line within it.

The 'combination principle' expressed by Walther Ritz in 1908, deduced from collected spectroscopic material that the frequency of every spectral line of a given element could be expressed as the difference between two terms—the so-called 'spectral terms'—each one of which depended on an integer number. By utilizing this principle all line series as they were discovered could be classified in a systematic way.

The regularities discovered by Balmer in the visible spectrum of hydrogen were found also in other spectral zones. Theodor Lyman (1874–1954) in 1906, while studying the hydrogen emission in the ultraviolet region, found that the series of lines emitted in this region could be represented by a formula similar to Balmer's; Friedrich Paschen (1865–1947) in 1908, and later, in 1922 the American astronomer Frank Parkhurst Brackett (1865–1953) and in 1924 the American August Herman Pfund (1879–1948) found similar results in the infrared.

All frequencies f of the different series could be derived using a unique formula:

$$c/\lambda = f = \text{const}(1/m^2 - 1/n^2)$$

where c is the velocity of light in vacuum and n and m are two integers that satisfy the following conditions:

$$m = 1, \quad n = 2,3,4,\ldots \quad \text{Lyman series in the ultraviolet}$$
$$m = 2, \quad n = 3,4,5,\ldots \quad \text{Balmer series in the visible}$$
$$m = 3, \quad n = 4,5,6,\ldots \quad \text{Paschen series (infrared)}$$
$$m = 4, \quad n = 5,6,7,\ldots \quad \text{Brackett series (infrared)}$$
$$m = 5, \quad n = 6,7,8,\ldots \quad \text{Pfund series (infrared).}$$

The effect of a magnetic field on the spectral lines

At the time the general behaviour of line spectra was explained, Pieter Zeeman (1865–1943) discovered in 1896 in Leyden, Holland, that a magnetic field is able to affect the frequencies of the spectral lines emitted by a gaseous source subjected to it.

Zeeman studied at Leyden University and earned his doctorate in 1893 by studying the Kerr effect which concerns the effect of a magnetic field on light polarization. In 1896, discussing his first experiments in a paper in the *Proceedings of the Royal Academy* of Amsterdam, he wrote that his discovery had been stimulated by that made by Faraday in 1854 concerning the influence of a magnetic field on the plane of polarization of linear polarized light (an effect similar to the Kerr effect which is today known as the Faraday effect). Already at that time, Faraday was convinced that light and magnetism were strictly linked. Maxwell tells us that Faraday dedicated his last experiments to the study of the influence of a magnetic field on light emitted by a source placed inside a magnetic field but that he did not report any result. Later, other researchers tried to repeat his experiments but failed.

Zeeman was a very meticulous experimentalist and thought that perhaps Faraday did not reach any conclusion because the effect was very small, and therefore planned the experiment with the greatest care, obtaining in 1896 a positive result. As already done by Faraday, Zeeman placed sodium chloride into a Bunsen burner which was itself positioned within the field region of an electromagnet and studied the spectral distribution of the emitted light using a very sensitive grating. He looked at the D-line of sodium (which in fact is a doublet, that is two lines very near to each other) and saw that when the electromagnet was switched on, the lines broadened. Initially he thought the effect was due to an influence of the magnetic field on the temperature and density of the vapour in the flame, but with successive experiments he found the phenomenon was effectively due to the influence of the magnetic field on the frequency of the D-lines of sodium.

Improved resolution later established that the effect consisted of the division of the emission line of zinc or cadmium into two or three lines, according to the orientation of the magnetic field with respect to the observation direction (figure 12).

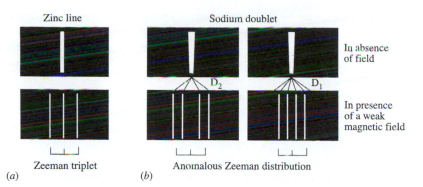

Figure 12. Examples of normal (for zinc) and anomalous (for the sodium doublet) Zeeman effect.

A short time before this discovery, H A Lorentz (1853–1928) had started to construct a theory of the behaviour of electrons, later published in his famous book *The Theory of Electrons* (Leipzig 1909), and he explained immediately the effect as being due to the electrons bound to atoms in a quasi-elastic manner.

Lorentz had also studied at Leyden University where he earned his doctorate in 1875. Twenty four years old, in 1877, he was appointed to the first chair of theoretical physics in the Netherlands, one of the first in Europe.

Lorentz had vast interests in physics and mathematics, but his most important contribution was the development of Maxwell's electromagnetic theory up to a point where the necessity of a radical change in the foundations of physics became clear and so gave the inspiration for Einstein's theory of relativity. To explain the negative result that in 1887 Albert Abraham Michelson (1852–1931) and Edward William Morley (1838–1923) obtained in an experiment to 'see' if light travels slower in the same direction of movement of the earth in space than at right angles to the Earth's surface, he assumed that material bodies in motion contract in the direction of motion and formalized in 1904 this hypothesis, that is today known under the name of 'Lorentz contraction', giving mathematical form to the transformations which describe it. The Lorentz transformations had a very important role in Einstein's theory which strengthened their basis theoretically.

In a series of papers published between 1892 and 1904, Lorentz built up an 'electron theory', which allowed him to explain a number of phenomena. He used his theory to explain the effect discovered by Zeeman, and the two researchers shared the physics Nobel prize in 1902 for the discovery and explanation of this important effect.

According to the Lorentz theory, light is emitted by atomic charged particles (electrons) whose motion is influenced by the magnetic field according to the laws of classical electromagnetism. From the frequency change produced by the magnetic field, Lorentz and Zeeman were able to determine the ratio between the charge and the mass of the particles which emitted light and the sign and value of the charge. Initially, they miscalculated and affirmed that the charge had positive sign, but then they corrected and gave it the correct negative sign. Meanwhile in Cambridge, J J Thomson, in experiments made in 1897, measured the ratio between the charge and the mass of the free electron and later in 1899 measured its charge separately. The values were identical to the ones found by Zeeman and Lorentz and demonstrated that electrons of whatever origin were identical.

The importance of the Zeeman effect in the theory of atomic structure must not be undervalued. The success reached by the Lorentz theory of the Zeeman effect showed that particles with the same charge/mass ratio of the electron were present in the atom and were responsible for the emission of the spectral lines. This was a direct confirmation that light emission occurred due to electrons.

After these first experiments, a number of other physicists, Preston, Runge and Paschen, and Landé, studied the magnetic splitting of spectral lines. The principal result of these investigations was that many lines, among which the sodium D-double, did not split into three lines, as predicted by the Lorentz theory but into a larger number of components (figure 12). This effect was called the 'anomalous Zeeman effect', and found its explanation only in 1925 when Uhlenbeck and Goudsmit introduced the 'electron spin'.

The first atom model

In conclusion we may say that in the first years of the 20th century a first, even if partial, answer was being given to the question of how light is emitted, and atoms with their electrical charges were identified as being responsible for this. However, how atoms were made, and through which processes light was emitted, nobody knew.

One problem concerned the number of electrons which resided in the atom. The first atomic models assumed this number to be very large. This hypothesis was supported by spectroscopic observations. Because spectral lines were assumed to be produced by electron oscillations the thousands of observed lines should be produced by thousands of electrons.

Thomson, to whom the discovery of the electron conferred a particular authority, produced in 1903–04 his model of the atom according to which the atom was a uniformly charged sphere with a positive charge in which electrons resided, dispersed like raisins in a plum pudding. The positive charge and the sum of the negative charges of all the electrons were equal in absolute value. Electrons were attracted towards the centre of the sphere and repelled each other according to Coulomb's law of electrical interactions. The normal state of the atom was reached when the system of these opposed forces was in equilibrium. If such atom was disturbed (or as physicists say, 'excited') by the collision with another atom or with a passing electron, its internal electrons started to vibrate around their equilibrium positions and light was emitted whose frequencies should be the ones measured spectroscopically. These frequencies could be calculated by applying the physical laws, and Thomson and his pupils made complex calculations to find, without success, the electron configurations which could give the right frequencies. The model was totally wrong, as we will see.

CHAPTER 3

BLACKBODY RADIATION

At the end of the 19th century, as we have seen, the conclusion was reached that light is an electromagnetic wave. However, at the same time at which the wave theory was gathering support, new phenomena were being discovered which contrasted with it. The study of the way in which a body absorbs or emits heat was among these phenomena. It was expected that the problem should attract a simple and immediate solution. However, this was not forthcoming, and when eventually it was found, it dealt the first blow to the wave theory of light.

Radiation and temperature

If we touch a body with our hand we may feel, if it has a high temperature, a warm sensation. This is the same sensation we experience if we are nearby without touching it, because the heat is transmitted through the air. However, even if we remove all the air which surrounds the body, heat transmission still takes place.

Today we know that the body transmits its heat, that is its energy, partly in the form of electromagnetic waves. The waves which transport the greatest part of the energy and are responsible for the warm sensation have the name of 'infrared radiation'. They have wavelengths which span almost the entire range between a millimetre and a thousandth of millimetre (micron) and are invisible to our eyes. However, the energy transported by visible light from the Sun over millions of kilometres may also transform into heat, as is well known by those who in summer like to become tanned by the sun's rays.

Friedrich Wilhelm Herschel (1738–1822), who was born in Hanover and later naturalized English, demonstrated early in the 19th century the heating effect of infrared radiation, showing that it may increase the temperature of a body. The experimental detection of infrared radiation improved notably in 1881 when Samuel Pierpont Langley (1834–1906) invented the so-called 'bolometer', an instrument able to measure the temperature of a platinum wire which was blackened with carbon, through the change of its electrical resistance.

To summarize, we may say that a hot body emits energy in the surrounding space partly in the form of radiation. As its temperature rises, the body irradiates an increasing amount of energy while the wavelength of the emitted radiation decreases going from the infrared region to the visible. At a temperature of about a thousand degrees, the body appears reddish; as the temperature increases further, it assumes a progressively intense colour passing from red to yellow, and eventually blue.

The exact relation which exists between the colour of a body and its temperature was discovered in the 19th century through a series of measurements and calculations initially all based on thermodynamics. Thermodynamics is that part of physics which focuses on the relation between work and the heat emitted or absorbed by bodies. It is founded on two fundamental principles: one states the impossibility to build an engine which indefinitely (that is cyclically) develops only work (that is, it establishes the principle of conservation of energy, and the impossibility of perpetual motion of the first kind), and the other one states the impossibility to have an engine that only takes heat from a source at constant temperature to transform it into work (that is, the impossibility of perpetual motion of the second kind). Note that all thermal machines take heat from a source at high temperature producing work; however, they also emit a portion of this heat at a lower temperature, for example to the external surroundings, and therefore not all the heat, but only a part of it, is transformed into work. From these two principles the most unexpected conclusions can be reached through pure logical reasoning in an absolutely rigorous and unexceptionable way because they require no specific model of the phenomenon to which they are applied.

At the beginning of the 19th century, researchers were greatly interested in the way a hot body emits radiation for a number of reasons, among which was their desire to create efficient light sources at a time when the illumination of cities with gas and electricity was just beginning. Moreover, the study of the light emitted from stars was at the time the only way to gain information about their nature.

Nobody imagined, however, that from the study of this problem one of the most profound and shocking revolutions of physics could arise: the revolution which leads to the 'quantum theory'. The final construction was the result of the efforts of many scientists in many different fields; here we limit our discussion to the considerations needed to understand lasers.

The blackbody

We may start by considering some results obtained by the German physicist Gustav Robert Kirchhoff.

Kirchhoff was born on 12 March 1824 in Königsberg, where he attended the University under the guidance of the physicist Franz Neumann (1798–1895). After earning his doctorate in 1847, Kirchhoff went to Berlin and became

Privatdozent (a title which gave him the right to teach in the university but not to have a salary; students paid a small amount of money directly to the teacher for his lectures) in that University a year later. In 1850 he was appointed extraordinary professor at Breslavia where he met the chemist Robert Wilhelm Bunsen (1811–1899) who some time later nominated him as physics professor at Heidelberg. In 1875 he was appointed to the theoretical physics chair at the University of Berlin where he died on 17 October 1887. He was nominated foreign member of the Italian Academy of Lincei in 1883.

Kirchhoff worked in nearly all fields of experimental and theoretical physics, obtaining everywhere results of fundamental importance. Besides the contributions that we will discuss here, he described the laws which allow us to derive the distribution of a current in an electrical network, he gave a noteworthy formulation of the two principles of thermodynamics, solved in a complete and rigorous way Maxwell's equations of electromagnetism, and strived to give a mathematical formulation for the Huygens principle.

In 1859, Kirchhoff started a conference at the Berlin Academy with the following words:

'A few weeks ago I had the honour of addressing the Academy with a memoir on some observations which seemed to me the most interesting as they allow us to draw conclusions on the chemical constitution of the solar atmosphere. Starting from these observations I have now derived, on the basis of rather simple theoretical considerations, a general theorem which, in view of its great importance, I allow myself to present to the Academy. It deals with a property of all bodies and refers to the emission and absorption of heat and light.'

Kirchhoff indeed did not lack modesty in presenting his results! He continued his lecture showing that bodies which emit radiation at some wavelength are able to absorb that same radiation, and for rays of the same wavelength at the same temperature, the ratio of the ability to emit radiation (the technical term is 'emissive power') to the ability to absorb it (absorptivity) is the same for all bodies and is independent of their nature and form.

Because this general result seemed to him very relevant, he emphasized the importance of making precise experimental measurements to verify his prediction and expressed the hope that the work would raise no special difficulties since 'all functions encountered so far which are independent of the nature of bodies are of simple structure'.

He then suggested for making these measurements the use of a body, which he named the 'blackbody', which was able to absorb all the radiation incident upon it. For this body, the absorptivity is unity and its emitting power becomes identical to the universal function he had introduced.

Although the ideal blackbody was an abstraction, Kirchhoff gave the necessary suggestion for a practical realization, by observing that it may be obtained by building a cavity with a small aperture of negligible diameter with respect to its dimensions. The hole would behave essentially as a blackbody; in fact every radiation incident on the hole may enter, i.e. is completely absorbed; after some time the interior radiation would reach equilibrium with the walls of the cavity at a temperature T and at this moment that radiation (which would be small with respect to that contained in the interior) would exit the hole as a sample of the characteristic radiation in the cavity.

A little later, in 1865, J Tyndall (1820–1893) published the results of some measurements of the ability of a body heated at two different temperatures to emit radiation. He had heated a blackened platinum wire, that is not exactly a blackbody. However, albeit imperfect, the measurements were used in 1879 by the Austrian physicist Josef Stefan (1835–1893) to state an empirical law according to which the quantity of energy emitted per unit surface area by a hot body is proportional to the fourth power of its absolute temperature.

The blackbody laws

Stefan's law attracted the attention of his pupil, Ludwig Boltzmann (1844–1906) who, by utilizing the principles of thermodynamics and electromagnetism, in 1884, delivered its theoretical proof using also a relation between the radiation pressure and the second law of thermodynamics, discovered several years before by the Italian physicist Adolfo Bartoli (1851–1896).

Ludwig Edward Boltzmann was born in Vienna during the night between Shrove Tuesday and Ash Wednesday, the son of a tax officer. Jokingly, he used to attribute his sudden temper changes between great happiness and deep depression to the fact he was born just at the death of a merry dance of Shrove Tuesday. He was a typical cyclothymiac, a short, robust man with curly hair; his girlfriend called him 'sweet dear fatty'. After earning his doctorate in Vienna in 1866, working under the guidance of Stefan to construct the kinetic theory of gases, he became his assistant and at the suggestion of Stefan, at only 25, was appointed to the mathematical physics chair at the Graz University.

While Boltzmann was studying for his doctorate, Stefan gave him Maxwell's papers on electrodynamics and an English grammar, telling him to learn English and read Maxwell. As a result Boltzmann wrote a paper on an electrodynamics problem. During his period in Graz, he wrote four fundamental papers on the statistical theory of gases, introducing all the concepts that later we will use in this book. But the most important result of this period was the introduction of an equation that facilitated the treatment of transport phenomena in gases (for example the heat or mass transport, that is diffusion, etc.) through statistical theory.

Independently from one another, Boltzmann and the American physicist Williard Gibbs (1839–1903) developed statistical mechanics, the science that provides the link between the microscopic world of atoms and molecules and the macroscopic world. At that time, chemists and physicists were very much interested in the problem of the 'reality' of atoms and molecules. For Boltz-mann they were real as much as the material objects that anybody could see and touch, but many people barely tolerated them as a useful concept to make calculations. Among the great opponents Wilhelm Ostwald (1853–1932), the author of a system called 'energetics' based on thermodynamics, was maintaining that all problems could be solved by reducing physics to the bare study of energy transformations. In 1895, Boltzmann, who was a fierce opponent of these concepts, arranged for Ostwald and his disciple, the mathematician Georg Helm (1851–1923), to be invited to the Gesellschaft Deutscher Natur-forscher und Arzte. In his lecture Helm maintained that the mechanical models, better the whole mechanics, had to disappear: the law of motion and also the behaviour of point masses had to be derived on the basis of simple energetic considerations. Boltzmann and others attacked him so strongly that Helm, afterwards, asked for and obtained a public apology. Even Ostwald was boycotted at the congress. Eventually, however, the anti-atomists disappeared, and in the end Ostwald converted.

In 1873 Boltzmann was awarded a mathematics chair in Vienna, but in 1876 he returned to Graz where he remained until 1890. During these years he focused on the radiation law. In Graz, however, he was not happy. In 1885 his mother died as did, in 1890, the first of his five sons. As a Dean of that University he had to deal also with the political problems caused by student anti-Habsburgic protests. In 1890 he accepted the chair in theoretical physics (his preferred discipline) in Munich and found new happiness there.

After the death of Stefan, in 1893, Boltzmann was invited to return to Vienna University where he remained until his death, with the exception of a short stay in Lipsia. During this time his theories on gases were criticized, as noted, and he had to struggle strenuously to defend them. This period was, however, very productive: he wrote four volumes on mechanics, electrody-namics and the theory of gases, and travelled greatly including to the United States of America, giving lectures and conferences. Towards the end of his career, his physical condition deteriorated and he became more and more depressed. In the summer of 1906 during a holiday in the Bay of Duino near Trieste in Italy, he hanged himself, while his wife and daughter were happily swimming.

The Stefan–Boltzmann law, that linked the temperature of a body to the amount of energy it is able to emit in the form of waves, was an important step in the foundation of the modern theory of radiating heat.

At that time the future Nobel prize-winner, Wilhelm Wien, was working at the Physikalisch-Technische Reichsanstalt (the Physico-Technical Institute at Berlin-Charlottenburg). The Institute had been founded in 1857 with the substantial help of the industrialist and inventor Werner von Siemens and was directed by Hermann Ludwig Ferdinand Helmholtz (1821–1894), the great German physiologist and theoretical physicist who invented the ophthalmoscope (1851), the instrument to inspect the interior of the eye, and the ophthalmometer to measure its curvature. As a theoretical physicist he made important contributions to thermodynamics by introducing the concept of free energy (the energy available to make work) and contributed to the discovery of the principle of conservation of energy.

Wilhelm Carl Werner Otto Fritz Franz Wien was born on 13 January 1864 in Gaffken, near Fischhausen, in East Prussia. After studying mathematics and physics at Göttingen, Berlin, Heidelberg, and Berlin again, earning a doctorate with a thesis assigned by Helmholtz concerning a diffraction problem, he worked for several years in a farm belonging to his father, until in 1890 Helmholtz called him as his assistant to the Physikalisch-Technische Reichsanstalt. His bright academic career developed in the Aachen, Giessen, Wurzburg and Munich Universities. He died in Munich on 30 August 1928. He has been one of the few physicists of the 20th century who worked as a specialist both in experimental and theoretical physics. His research on the blackbody earned him the Nobel prize for physics in 1911 'for his discoveries regarding the laws governing the radiation of heat'. He worked in thermodynamics and performed pioneering experimental studies on the electrical and magnetic deflection of canal and cathode rays (the rays produced in electrical discharges in gases) contributing to the discovery of the electron.

The experimental study of the distribution of the frequencies emitted by a blackbody at a given temperature showed that the intensity has a maximum at a wavelength which changes with changing temperature, and becomes shorter and shorter as temperature increases (figure 13). In 1893, on the basis of thermodynamic reasoning, Wien presented his explanation of this result.

The law, called the 'displacement law', establishes that the product of the wavelength at which the maximum of radiation occurs times the absolute temperature of the body, is a constant. By introducing some very general hypotheses on how radiation could be emitted by a body—hypotheses based on an idea of the Russian Vladimir A Michelson (1860–1927) who in 1887 proposed an explanation for the continuity of the distribution of energy in the spectra of solids on the basis of atomic vibrations—Wien assumed that blackbody radiation was produced by oscillators of atomic dimensions, eventually formulating in 1896 a radiation distribution law for the blackbody. The law provided some indications of the dependence of intensity on wavelength at a given temperature, and gave an adequate account of all experimental data available at the time. These data, however, due to the absence of good detectors for long wavelength radiation, did not extend to wavelengths longer than a few microns.

Max Planck and the blackbody law
While the Physikalisch-Technische Reichsanstalt was becoming increasingly involved in the preparations for absolute measurements of blackbody radiation, in June 1896, Wien left Berlin to take a professorship at Technische Hochschule of Aachen. Fortunately Max Planck, who had succeeded

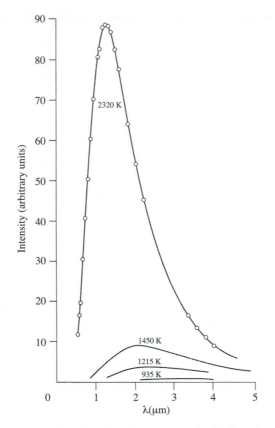

Figure 13. The curves, given for the indicated temperature (in K) show, in arbitrary units, the intensity of the emitted radiation as a function of the wavelength (λ) expressed in microns (μm). Simple inspection shows that, by increasing the temperature, the emitted intensity increases and its maximum value shifts towards shorter wavelengths.

Gustav Kirchhoff as professor of theoretical physics at the Berlin University, became 'resident theoretician' for the experimentalists of the Institute who worked on blackbody radiation.

Max Karl Ernst Ludwig Planck (who was born on 18 April 1858) was the son of a law professor at Kiel, who later transferred to Munich where Max attended the University. He later justified his choice to study physics: 'The external world is something independent from man, something of absolute, and the search for the laws that apply to this absolute looked to me as the more sublime purpose in my life'. Later he recollected that, when at school, he learned the principle of the conservation of energy 'my mind absorbed avidly, as a revelation, the first law to me known which could have a universal absolute validity, independent from any human agent'. Thermodynamics remained his preferred subject and his research in this field, already started with his dissertation, procured him, in 1889, the appointment of professor at the Berlin University, as a successor of Kirchhoff, who had died some time before.

When, after obtaining the notable results of his scientific career some of which we will discuss shortly, Planck retired, he was the most prominent personality in German science and an important figure in the relations with Nazism. He had a cautiously cooperative attitude, and whilst he did not publicly protest against the prosecution of Jewish scientists, considering it the fruit of a temporarily foolish administration, in 1933 he discussed the issue with Hitler, observing that the racial laws had impoverished German science. The answer was that science could be passed over for a few years.

In his private life, Planck was very unfortunate. During the First World War, his son Karl died, and later in 1917 and 1919 his twin daughters Grete and Emma both died in childbirth. During the Second World War he watched over the destruction of his country, had his home razed by bombs, and his last son Erwin was executed in 1945 for complicity in the 1944 attempt against Hitler's life. He died in Gottingen in 1947.

At Berlin the Physikalisch-Technische Reichsanstalt was, as we said, involved in precision measurements of universal constants and functions, and particularly the blackbody distribution function and relative constants, and Planck was in close communication with the physicists who worked on it. He decided to study the problem, trying to find a justification of Wien's law by utilizing only thermodynamic and electrodynamics ideas, and publishing five papers on the subject between 1897 and 1899. The basis of his method consisted of assuming that the cavity walls could be represented as an ensemble of harmonic oscillators, that is charges in oscillatory motion (that H Hertz had demonstrated a short time earlier were able to emit electromagnetic waves), and that an equilibrium was established in the cavity between the emitted and absorbed radiation.

After a discussion with Boltzmann who criticized and corrected a formulation error, Planck obtained a simple relation between the average energy of the Hertzian oscillators and the energy distribution in the blackbody, based on these hypotheses. In May 1899 he presented his results to the Prussian Academy justifying Wien's law, which was deduced considering a particular thermodynamic property of the oscillators, that is their entropy. In fact Planck, as he explained later, did not deduce Wien's law from an independent calculation of the entropy of the oscillators but used Wien's law to obtain the form of entropy that oscillators should have, verifying that this expression was not in contrast to any thermodynamic rule. He did so because at the time Wien's law seemed to explain quite well the available experimental results.

Rayleigh's law

At the end of 1899, more precise measurements extended to longer wavelengths were performed, which showed that in this region Wien's law was no longer valid. In June of that same year, Lord Rayleigh, who was born John William Strutt (1842–1919) published a deduction of the distribution law derived on a purely electromagnetic basis, with which he demonstrated that the radiation intensity was proportional to temperature.

> The Strutts originated from John Strutt (died in 1694), a corn miller in Essex whose offspring became members of Parliament. Joseph Holden Strutt distinguished himself in the war against Napoleon and was rewarded by King George III with a barony. Rather than give up his Commons seat in Parliament by becoming a peer, he declined the honour, and requested instead that his wife, Lady Charlotte, a daughter of the first Duke of Leinster, be created the first Baroness. The name of the small town of Rayleigh was chosen as the territorial title, simply because it was a melodious Essex name. At the death of his mother, John James Strutt (1796–1873) became in 1836 the second Baron, and the eldest of his six sons, John William Strutt was therefore the third Baron.

The initial education of the young John William Strutt was frequently interrupted by ill health. After being at Eton, Harrow and a school in Torquay, in October 1861 he went to Trinity College, Cambridge. At that time Cambridge University in general, and Trinity College in particular, dominated the production of British mathematicians and natural scientists (physicists), and boasted as early graduates Bacon, Newton and Cavendish. In those days, a student attended college, heard lectures and studied, but the mark of his achievement and even his future career depended on his standing in the Natural Sciences or Mathematical Tripos examinations. The top man was, until 1912, called the Senior Wrangler, the next, Second Wrangler, and so on. The man who had the misfortune to be at the bottom of the list received the 'Wooden Spoon' and bore this humiliation for the rest of his life. The results were published in the newspapers, and served as a basis for awarding fellowships and even professorships. The system was, however, criticized, and William Thomson (1824–1907) who was Second Wrangler described it as 'miserable', and also J J Thomson, who in 1880 was second to Joseph Larmor (1857–1942), criticized it. At that time in the United Kingdom, of the 37 chairs, 18 were occupied by Wranglers. Maxwell, J J Thomson and Kelvin were all Wranglers. The importance of Wranglers in late Victorian physics is evidenced by the candidature, to the mathematical physics chair in 1881 at Manchester University, of the English naturalized German physicist and astronomer Arthur Schuster (1851–1934). Schuster presented certificates of his mathematical abilities from masters like Kirchhoff, but from nobody with a Cambridge background. A friend advised him in due time that surely his application had been unsuccessful, and advised him to obtain a certificate from a Wrangler, which in Manchester would be worth the references of more than a thousand Kirchhoffs. Fortunately Schuster knew someone, and in due course was appointed to the chair.

In January 1865, J W Strutt took his degree and became Senior Wrangler in the Tripos being recognized throughout Britain as a mathematical physicist of great ability and promise which allowed him to win a fellowship at Trinity College.

The scientific world was living through a stimulating period. When Strutt was in school, Joule's experiments on the conservation of energy had just received general acceptance, and William Thomson (later Lord Kelvin) had given his formulation of the second law of thermodynamics. Maxwell applied the laws of probability to the kinetic theory of gases about the time Strutt entered University and when he came out, Maxwell was communicating to the Royal Society his paper on 'A dynamical theory of the electromagnetic theory', the mathematician G Green (1793–1841) and the physicist G G Stokes (1819–1903) had put the undulatory theory of light on a sound basis, and the physicist J B L Foucault (1819–1868) had shown experimentally that light propagated more slowly through water than in air. Michael Faraday was still alive, albeit old. Experimentally, Bunsen and Kirchhoff had introduced spectral analysis, photography was still in its infancy, and the first observations had been made of electrical discharges in gases. In the laboratory, a vacuum was achieved only after hours of hand operation of special pumps, and electricity came from manually-operated generators or batteries.

Strutt's first interests were acoustics and the physiology of hearing, displaying the great physical aptitude and mathematical skilfulness that may be found in his book *The Theory of Sound*, published in 1877 and still important today. In 1871 he married Evelyn Balfour, the sister of his college friend Arthur Balfour, and settled on the family estate at Terling, about 70 km north-east of London, where he converted some of the rooms to a private laboratory. Here he commenced a series of experimental studies in acoustics and optics. Being interested in photography since a child, he described a technique for colour photography (1887) which was later, in 1891, successfully carried out by G Lippmann (1845–1921) who was awarded in 1908 the physics Nobel prize for 'his method of reproducing colours photographically based on the phenomenon of interference'. By studying diffraction gratings, Strutt introduced his famous criterion for the resolving power which today is known under his name (Rayleigh's criterion). In conducting his early experiments on colour vision he was led to study the blue colour of the sky, and correctly attributed it to the scattering of light from air molecules, deriving the quantitative relationship which describes it.

Until the second half of the 19th century, there was no teaching laboratory of 'practical physics' within university courses, and the great experimenters had learned experimental techniques in private laboratories, or by helping an older master in his work. In 1850, William Thomson (Lord Kelvin) cleaned out an old wine cellar in the college basement at Glasgow and assembled a student laboratory for research into electricity. Other laboratories were prepared and in 1869 a committee at Cambridge recommended that a special professorship be set up to teach and demonstrate experimental physics. The next year, the Chancellor of the University, the seventh Duke of Devonshire (William Cavendish) supported the construction of a research laboratory, the Cavendish Laboratory, and a chair which was occupied by Maxwell until his death in 1879. In December 1879 Lord Rayleigh was invited to cover the chair and when he retired at Terling in 1884, the chair was assigned to his pupil J J Thomson.

In 1896 Rayleigh accepted the position of Secretary of the Royal Society, and in 1897 succeeded John Tyndall as Professor of Natural Philosophy at the Royal Institution, which he held until 1905. To this period belongs the discovery, made together with William Ramsey (1852–1916) of London University, of the gas argon for which Rayleigh was awarded the Nobel prize for physics in 1904, and Ramsay the Nobel prize for chemistry in the same year.

The last years from 1895 to 1919 were years of honours and glory. When, in 1899, he became interested in the blackbody problem he was at The Royal Institution.

At this point it is worth explaining in some more detail the reasoning made by Rayleigh, because some concepts will be useful later.

The kinetic heat theory, developed principally by Ludwig Boltzmann, James Clerk Maxwell and Josiah Willard Gibbs, considers heat as the result of the casual motion of the many atoms or molecules from which all material bodies are formed. Because it would be futile to attempt to follow the motion of each single atom or molecule which takes part in thermal motion, the mathematical description of thermal phenomena must necessarily make recourse to the statistical method. To explain the macroscopic behaviour, for example of a gas, mean quantities are considered by considering a large number of molecules.

One of the fundamental principles of such a method is the so-called equipartition theorem, derived by Maxwell and which may be mathematically derived from Newton's principles of mechanics. The theorem states that 'the total energy contained in a system with a great number of particles that exchange energy among them by means of reciprocal collisions, equally distributes (in the mean) among all particles'. If, for example, E is the total energy and N is the number of particles, on average every particle has an energy E/N.

Although the equipartition principle regulates the energy distribution among a great number of particles, the velocity and energy of an individual particle may depart from the mean value, fluctuating in a statistical way around this value. This means that if the mean energy has a certain value, some molecules may have larger energies and other smaller energies. These different values are called fluctuations. If we represent mathematically the fluctuations, for example of the velocity of molecules in a gas, we obtain curves that show the relative number of particles having a certain velocity for every temperature, velocities larger or lower than the mean value.

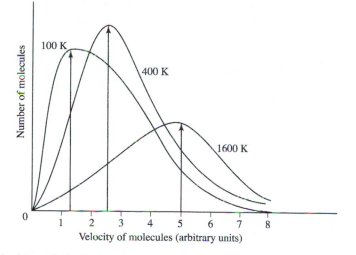

Figure 14. Maxwell distribution: the number of molecules with a given velocity is given as a function of velocity for three different temperature values, 100, 400 and 1600 K. Because the number of molecules inside the vessel does not change, the areas under the three curves are equal. The mean velocity of molecules (shown by the arrows) increases proportionally to the square root of the absolute temperature.

These curves, which were derived for the first time by Maxwell and bear his name, are represented in figure 14 for three different temperatures of a gas.

The use of the statistical method in the study of the thermal motion of molecules explained very well the thermal properties of material bodies, especially in the case of gases.

The idea of Lord Rayleigh was to extend the statistical method to thermal radiation. By studying the intensity distribution of light emitted at different frequencies as a function of temperature, one obtains curves such as those in figure 13 in which the distribution is given for three different temperatures. These curves, compared with those of figure 14, show a notable similarity: while in figure 14 the temperature increase shifts the maximum of a curve towards higher velocities, in figure 13 the maximum shifts towards higher radiation frequencies. This fact encouraged Rayleigh to apply to thermal radiation the same equipartition principle that had so much success in the gas case: that is to assume that the total radiative energy be equally distributed among all the possible vibration frequencies (the so-called modes). The result seemed correct from the classical point of view. Rayleigh, however, made a small error in counting the number of modes, which later in 1906 was corrected by the physicist, astronomer and mathematician James Jeans (1877–1946), so that the formula is today remembered as the Rayleigh–Jeans law. At large wavelengths the formula explained very well the experimental results; it failed, however, at short wavelengths where strange results were obtained. The trouble was that, in spite of all the similarities

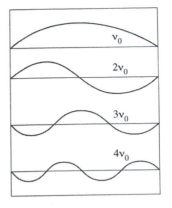

Figure 15. Vibrations of a cord. In the upper part of the figure is shown the fundamental vibration and, moving down, its harmonics.

between the gas composed of single molecules and thermal radiation constituted by electromagnetic vibrations, a substantial difference existed: while the number of gas molecules in a given closed space is always finite, even if it is very great, the number of possible electromagnetic vibrations (modes) in the same space is always infinite.

To understand this fact we may consider the simple case of the undulatory motion which takes place in a single direction (unidimensional) represented by the motion of a cord fixed at both its ends. Because the extremities of the cord cannot move, the only possible vibrations are of the kind shown in figure 15 which correspond in musical language to the fundamental note and the various harmonics: within the total length of the cord may exist a half-wavelength, or two half-wavelengths, or three, or ten, or a thousand, or any integer number of half-wavelengths. The corresponding vibration frequencies are factors of two, three, ten, or a thousand times that of the frequency of the fundamental note.

In the case of stationary waves inside a box in three dimensions, as for example a cube, the situation is the same even though a little more complex, but results in the same way in an unlimited number of different vibrations, with wavelengths shorter and shorter and corresponding frequencies higher and higher. If we apply the equipartition energy principle, by calling E the total energy available in the box, the energy E divided by the total number of modes will correspond, because the mode number is infinite, to the energy of any single vibration and this will be an infinitely small quantity of energy! This conclusion is clearly absurd, and becomes even more unlikely if we apply it to Kirchhoff's blackbody. If we allow a small quantity of radiation at some wavelength, for example in the red, to enter the cavity, it will start to interact with the walls of the box and shall be distributed among the infinite vibration modes contained in the cavity, that is among infinite frequencies ranging from those lower than red to those higher than red, that is violet, x-rays, γ-rays and so on. This paradoxical result was called the 'ultraviolet catastrophe'. According to this analysis, the open door of our kitchen oven should become a source of x-rays and γ-rays, like an atomic bomb!

Rayleigh's paper, published in June 1900, was only two pages long but showed in a clear and unequivocal way the inevitable result that was obtained by applying classical statistical mechanics to the radiation problem. Neither Planck, nor his experimental colleagues H Rubens (1865–1922) and F Kurlbaum (1857–1927), took Rayleigh's paper very seriously. The distribution law proposed by Rayleigh if contrasted with the experimental results was in disagreement except for long wavelengths. Therefore, at first it was dismissed along with several other laws that had been proposed on the basis of various hypotheses.

Planck's law
The theoretical situation was as just described, when on Sunday 7 October 1900, H Rubens during a visit paid with his wife to Planck, told him about measurements made in collaboration with F Kurlbaum at the Berlin Institute, extended to wavelengths up to 50 µm, which showed a marked deviation from the predictions of Wien's law, and were in agreement with the new Rayleigh formula. A public presentation of these results had to be given at the 19 October 1900 session of the Deutsche Physikalische Gesellschaft. Before the meeting Planck tried to modify his expression for the entropy of the oscillators so as to take into account the new results. Still arguing on the grounds of thermodynamic considerations, he derived the distribution law that is today known under his name. The same night he sent a postcard to Rubens with the new formula, which was received the following morning. One or two days later Rubens came back to Planck and showed him that the experimental results perfectly fitted the new formula. On 19 October at the meeting of the Deutsche Physikalische Gesellschaft, Kurlbaum presented the experiments performed together with Rubens, and in the lively discussion that followed, Planck presented his new formula in a comment with the title 'On an improvement of Wien's radiation law'. 'The same day in which I formulated the law I started to devolve myself to the task of dress it with a correct physical meaning' said Planck later, who after a few weeks of the most strenuous work of his life, on 14 December, again at the Deutsche Physikalische Gesellschaft, was able to explain the physical hypotheses which supported the law.

In his lecture Planck stated that, according to some rather complex calculations he had performed, a remedy could be found for the paradoxical conclusions obtained by Rayleigh, and the danger of the ultraviolet catastrophe could be avoided if the postulate was made that the energy E of electromagnetic waves (visible light included) may exist only in the form of some packet, with the energy content of every packet directly proportional to the corresponding frequency f:

> '...we consider—and this is the most important point of the entire calculation—E as being composed of a completely definite number of finite, equal parts, and make use for that purpose of the natural constant h.... This constant, when multiplied by the common frequency f of the resonators, yields the energy element ee...; and by dividing E by ee we obtain...the number of energy elements which have to be distributed among the N resonators'.

This hypothesis, known as the quantum theory, assumes that energy may be emitted only in discrete quantities, or packets, and not in continuously variable quantities. The minimum energy an oscillator may emit at the frequency f is the product of the frequency times a universal constant, that Planck indicated with the letter h, and is today known as Planck's constant (the action constant).

Planck obtained his interpretation of the blackbody law before the middle of November 1900, but presented his results for the first time at the meeting of the Deutsche Physikalische Gesellschaft in Berlin only on 14 December. The great mathematician and physicist A J W Sommerfeld (1868–1951) called this day 'the birthday of quantum theory'. He referred in particular to the fact that Planck had considered to be the 'most essential point' of his theory the hypothesis that energy is distributed among the cavity resonators only by integral multiples of finite energy elements.

More than 30 years later, in a letter to his friend the physicist, specialist in optics and spectroscopy, Robert Williams Wood (1868–1955) on 7 October 1931, Planck gave this justification:

> 'In short, I can characterise the whole procedure as an act of despair, since, by nature I am peaceable and opposed to doubtful adventures. However, I had already fought for 6 years (since 1894) with the problem of equilibrium between radiation and matter without arriving at any successful result. I was aware that this problem was of fundamental importance in physics, and I knew the formula describing the energy distribution in the normal spectrum (i.e. the spectrum of a blackbody); hence a theoretical interpretation had to be found at any price, however, high it might be.'

Paradoxically this revolutionary hypothesis by Planck was not immediately recognized, and contemporary physicists did not understand that a new physics had been born. Planck was himself unaware of the revolution he was starting, considering at first the energy quantization to be little more than a simple mathematical modification useful for making calculations. He did not think that energy was really concentrated in discrete quanta; being a profoundly conservative man, for some years he limited his thoughts concerning his energy quantization theory simply as a convenient hypothesis that allowed Boltzmann statistics to be applied to the radiation problem.

Also the physicists of the first years of the 20th century used the black-body formula as an empirical formula and Planck himself, after having introduced the concept of quantization, tried to limit it and produced two successive modifications of his theory in which he succeeded in obtaining the same formula without the need for the assumption that absorption processes involve the exchange of energy in discrete quantities, that is, energy quanta (1914). The scientific community took several years to recognize his contribution, awarding him the Nobel prize in physics only in 1918 'in recognition of the services he rendered to the advancement of Physics by his discovery of energy quanta'.

Among the first to point out that something was not quite right was Lord Rayleigh in 1905 who, coming back to his formula of 1900, observed that Planck's formula reduced to his one in the limit of low frequencies and concluded

> 'A critical comparison of the two processes [i.e. his own and Planck's] would be of interest, but not having succeeded in following Planck's reasoning, I am unable to undertake it. As applying to all wave lengths, his formula would have the greater value if satisfactorily established. On the other hand, the reasoning which leads to [my equation] is very simple, and this formula appears to me to be a necessary consequence of the law of equipartition as laid down by Boltzmann and Maxwell. My difficulty is to understand how another process, also based upon Boltzmann's ideas, can lead to a different result.'

So Rayleigh pointed out the existence of the introduction of a new concept that is usually called the 'crisis of classical physics', because effectively in the context of that physics the Rayleigh deduction appeared correct and inevitable.

At this time the genial considerations of an unknown employee of the Patent Office in Bern, Switzerland, secured the theoretical basis for the understanding of light emission and absorption phenomena. This obscure employee was Albert Einstein. As we will see, Einstein accepted completely the consequences of the quantization hypothesis and assumed that radiation behaves as if it consisted of energy quanta that manifest themselves not only in the emission and absorption processes but exist independently as particles in vacuum. Before discussing his fundamental contributions, however, we need to illustrate another important revolution that occurred concerning the discovery of the atomic structure and of its role in light emission.

CHAPTER 4

THE RUTHERFORD–BOHR ATOM

The concept of an atom, as a complex system containing in its interior both negative charges (electrons) and positive charges (necessary to neutralize the negative charges and make the atom electrically neutral) was evolving, as we have seen, in the years between the 19th and 20th centuries. In 1911, thanks to a fundamental experiment performed by Rutherford, a model was developed which, with suitable modifications, we still use.

Rutherford and the planetary atom

Ernest Rutherford (1871–1937) was born at Spring Grove, a locality near Nelson on New Zealand's South Island, in a family of Scottish origin. His mother was a schoolteacher who played pianoforte, a rather unusual thing in New Zealand of that time. His father, a formerly energetic and ingenious farmer, had founded a prosperous business in linen production, and the young Ernest, his six brothers and five sisters (three of whom died in their childhood) lived in a farming community.

When 10 years old, Ernest read a popular physics book by Balfour Stewart (1828–1887) and, as happened to many other physicists in similar cases, remained fascinated by the subject. After secondary school and college, coming first in all examinations in English, Latin, history, mathematics, physics and chemistry, in 1889 he was awarded a University scholarship at the University of New Zealand where he earned his degree presenting a dissertation on the magnetization of iron produced by high frequency electrical discharges, winning, in 1894, a scholarship which allowed him to continue his studies in England. The story goes that he received the news while digging potatoes and exclaimed 'That's the last potato I'll dig in my life' after which, having borrowed the money for the ticket, he left for Cambridge, in 1895, where he was admitted as a research student to the famous Cavendish Laboratory directed by J J Thomson, the discoverer of the electron.

Cambridge University had just recently decided to focus more on experimental research, opening the laboratories also to students coming from other universities, and among these the first was Rutherford. His fame spread rapidly among his school fellows and one of them wrote:

'We've got a rabbit here from the Antipodes, and he's burrowing mighty deeply'.

At Cambridge, Rutherford continued his studies on magnetism, obtaining interesting results on the transmission and detection of electromagnetic waves; then, after the discovery of x-rays (in 1895) by W Roentgen, with the enthusiasm and energy which were the salient features of his personality he joined Thomson in his research on x-rays, and later on radioactivity (1896). In this field he provided fundamental contributions, working first at Cambridge and, after 1898, at Montreal in Canada where he was appointed professor of physics at McGill University, by developing the idea that radioactivity is actually the destruction of the original atoms and their transformation into atoms of different elements; with this disintegration theory he unravelled the nature of radioactive phenomena. The theory received full support by the experiments he performed together with his young collaborator, the chemist Frederick Soddy (1877–1956), who was rewarded with the Nobel prize in chemistry in 1921 'for his contributions to our knowledge of the chemistry of radioactive substances, and his investigations into the origin and nature of isotopes'; in particular the nature of some radiations emitted by the radioactive bodies that Rutherford had discovered and called alpha and beta rays was clarified (helium nuclei the alpha, electrons the beta). In 1907 he returned to the UK, succeeding Arthur Schuster as professor of physics in Manchester, and the following year was awarded the Nobel prize in chemistry 'for his investigations into the disintegration of the elements, and the chemistry of radioactive substances'. Eventually, in 1911, by studying the diffusion of alpha particles though a solid target he proposed the planetary interpretation of the atom. In 1919 he announced the first artificial nuclear disintegration, and succeeded J J Thomson as director of the Cavendish Laboratory. He was knighted in 1914, and in 1932 was created first Baron Rutherford of Nelson. He was the president of the Royal Society from 1925 to 1930.

When he died unexpectedly from a strangulated hernia, his ashes were buried in the nave of Westminster Abbey, just west of Newton's tomb and by that of Lord Kelvin, in the presence of the representatives of the King and government.

Rutherford, who was considered a dominant personality in the development of physics in that period, was a man of volcanic energies, great enthusiasm, enormous work capacity and solid common sense. A collaborator of his said that he was a man 'not smart but simply great'. False modesty was unknown to him.

In 1911 he postulated a model of the atom in which finally the correct distribution of negative (electrons) and positive charges was given. With an experiment that has become a classic in the history of physics he demonstrated that a strong positive charge concentration resides in the central region of each atom where is concentrated also most of the atomic mass.

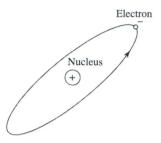

Figure 16. Artist's view of the Rutherford atom. The example refers to hydrogen. The electron (negative charge) turns around the nucleus (positive charge) like the Earth turns around the Sun.

This central region which is at least 100 000 times smaller (in linear dimensions) than the whole atom is today indicated with the name of atomic nucleus. The negative charge which surrounds the nucleus is made by electrons which revolve around it under the action of electric forces of mutual attraction. Because the atom as a whole is electrically neutral, the total negative charge of electrons which turn around the nucleus must be equal to the positive charge of the nucleus (figure 16).

Atoms of different elements contain different numbers of electrons which revolve around the nucleus. This result was reached gradually, starting from Mendeleev's discovery that the known elements can be ordered according to a progression of increasing weight in a table in such a way that the ones with similar chemical properties are all in a same column. Later (1913) the English physicist H G J Moseley (1887–1915), who died very young in Gallipoli during the First World War, performed a series of measurements on the scattering of x-rays from atoms which allowed him to determine the number of electrons actually contained in the atoms, showing that the sequence of elements in Mendeleev's table was generated moving from one element to the next by adding an electron. The conclusion was then reached that the number of electrons was much smaller than imagined, that hydrogen is the simplest atom with a single electron, helium has two electrons and so on, until the heaviest known element at the time, uranium, was reached whose atoms contain 92 electrons (today atoms which contain up to 118 electrons have been artificially produced).

We may say that this model of the atom is similar to the system of planets which turn around the Sun under the action of gravitational forces with an important difference that must not be neglected. The electrons which revolve around the nucleus carry an electric charge and therefore are obliged, according to Maxwell's laws of electromagnetism, to emit electromagnetic waves like the antenna of a broadcasting station. But because these 'atomic antenna' are much smaller, the electromagnetic waves emitted by atoms are milliards of times smaller than those emitted by a common antenna. These waves occur in the visible region and their emission makes the body luminous.

The electrons which turn around the nucleus in the Rutherford model should therefore emit light waves and because these waves transport energy they should constantly lose kinetic energy as a consequence of their emission. It is easy to calculate that, if this was true, all atomic electrons would lose completely their kinetic energy in a negligibly small fraction of second, and would fall on the nuclear surface.

However, observation shows that this does not take place, and the atomic electrons revolve perpetually around the nucleus, at a relatively large distance from it. In addition to this contradiction regarding the fundamental nature of the atom, a number of other inconsistencies also existed between the theoretical predictions and the experimental results. For example, experience says that atoms emit light of only certain colours, or wavelengths (the spectral lines that we discussed in chapter 2) while the electron motion in the Rutherford model would tend to an emission of light of all colours (that is of all wavelengths).

Niels Bohr

The team of young people who were around Rutherford in Manchester was mostly experimental physicists, and similarly Rutherford, albeit too intelligent not to appreciate the importance of the theory, was essentially an experimentalist. 'When a young man in my lab uses the word universe', he roared once, 'I say it is time for him to go'. 'And why do you trust Bohr?' he was asked. 'Oh', he answered, 'but he is a football player!'.

The chair in Manchester, one of the principal provincial English universities, passed to Rutherford when the spectroscopist Sir Arthur Shuster decided to retire, provided Rutherford succeeded him. Of German origin, Schuster had inherited a fortune that he partly used to provide his Institute with a very beautiful laboratory, supporting also a scholarship for a theoretical physicist that was successively assigned to H Bateman (1882–1946), C G Darwin and to a young Danish physicist, Niels Bohr (1885–1962).

Niels was born in Copenhagen to a well-off family. His father was a well-known physiology professor and his mother originated from a cultured family of English Jewish bankers.

At the time, Denmark was culturally midway between the English and German traditions, allowing a fortunate synthesis of English experimental science with the more formal, theoretical approach of German universities. In many ways Bohr's temperament combined the British influences originated by the Locke empiricism of common sense, with the typical German heritage of Kantian attitudes with subjective and objective aspects of experience.

Bohr had one older sister, Jenny, and a brother, Harald (1887–1951), one and a half years his junior. The relationship between Niels and Harald

67

was remarkable and had an important influence on Niels' method of working. From childhood the two brothers learned to express and sharpen their ideas in vigorous dialogue, thereby developing an incredibly interdependent dialectical rapport with each other. Their continuous dialogue conditioned Bohr to need to work out his ideas by discussing them with another. The relationship with Harald, who later became a famous mathematician and director of the Institute of Mathematics next door to Niels' Institute for Theoretical Physics, provided the mathematical knowledge necessary for Niels' work.

In the spring of 1911, Niels finished and defended his doctoral dissertation on the electron theory of metals. At the turn of the century, several important physicists, on the basis of the evidence of the ubiquity of the electron, given by Joseph John Thomson, and by the Hendrik Antoon Lorentz's theory of electron behaviour, tried to explain all physical phenomena as consequences of the interactions of electrons among themselves and with molecules.

The first success was obtained with the theory of metals. Thomson, Lorentz, Paul Drude (1863–1906) and others obtained a promising agreement with experiment on the assumption that electrons move through metals as do molecules through a perfect gas. In 1900, Drude deduced that the ratio of thermal conductivity to electric conductivity should be the same for all metals, and directly proportional to the absolute temperature. His expression was, however, only one half of that found by experiment. Lorentz in 1905, treating the free electrons in a metal by the statistical methods used for gases, obtained results in greater disagreement with experiment than the ones of Drude. Even the radiation emitted as heat from metals had been calculated by Lorentz in 1903, and Paul Langevin (1872–1946) in 1905 had presented a theory of magnetic behaviour.

Bohr in his dissertation considered all these different problems and concluded that the electron theory of metals could be modified so as to give results in agreement with experiment when the internal structure of atoms was not involved. The problem of radiation, and the explanation of the magnetic behaviour, needed instead new radical hypotheses of which, however, he had no precise idea.

At that time a doctoral graduate was used to doing postdoctoral research outside of Denmark, and quite predictably—having discussed in his dissertation just the behaviour of electrons in metals—Bohr chose to go to Cambridge to work with J J Thomson.

The first meeting with Thomson did not establish a good relationship between them. Bohr entered Thomson's office with one book, opened it and politely said: 'This point is wrong'. We should note that at the time Bohr did not speak English very well, and had, therefore, condensed his thought into short sentences. In any case in October 1911 he wrote to his brother Harald:

'... Thomson has so far not been easy to deal with as I thought the first day. He is an excellent man, incredibly clever and full of imagination (you should hear one of his elementary lectures) and extremely friendly; but he is so immensely busy with so many things, and he is so absorbed in his work that it is very difficult to get to talk to him. He has not yet had time to read my paper [his thesis], and I do not know yet if he will accept my criticism.'

Effectively, Thomson had ceased to work on the theory of metals and moreover was temperamentally averse to the close collaboration and constant conversation that Bohr needed in order to develop ideas. Nevertheless, whilst at Cambridge, Bohr familiarized himself with Thomson's work on atomic models and reached an understanding of their fundamental inadequacy, but overall he was not very satisfied.

Bohr and Rutherford's atom
The year before Bohr's arrival in England, Rutherford had made the discovery of the atomic nucleus, and in the autumn of 1911 the two men met and undoubtedly both were impressed with each other, for by March 1912 Bohr left Cambridge for Manchester with the intention of working experimentally on radioactivity. Here he started to study the slowing down of alpha particles in their passage through matter, then after a few weeks he began to concentrate on the theoretical aspects, by considering the interaction of alpha particles with the electrons of the atom. So he refined a theory of one of Rutherford's collaborators, Charles Galton Darwin (1887–1962)—the grandson of the father of the theory of evolution Charles Robert—whom Bohr used to call 'the grandson of the real Darwin'. Charles Galton Darwin had assumed that the loss of energy of alpha particles travelling in matter was due entirely to collisions between the alpha particles and the electrons of atoms. In his model, Darwin assumed electrons could be treated as free (not subjected to any force) particles and Bohr refined this model by treating the electrons that surround the nucleus as 'harmonic oscillators', that is by assuming that they were linked to the nucleus by elastic forces, and should have their energy quantized, according to the Planck quantum rules. Bohr finished work on the argument only after he left Manchester and the result was published in 1913. The study stimulated Bohr's interest in the problem of atomic structure. Already in Manchester he had started to sketch ideas on the stability of the atom; completely new ideas of which he had given a preliminary report to Rutherford.

Bohr left Manchester on 24 July 1912 to return to Denmark where, on 1 August, he married Margrethe Norlund. During the spring and summer he realized that Rutherford's discovery of the atomic nucleus was a cornerstone for the construction of a model of the atom, and that no Rutherford atom could be represented as a mechanically stable system according to the laws of classical physics. At the same time he convinced himself that the action quantum should play a role in the development of any atomic theory.

At the beginning of 1913, H M Hansen (1886–1956)—a young man from Copenhagen who had carried out experimental research on spectra in Göttingen—directed his attention towards the discovery, made by Balmer in 1885, that the light emitted by hydrogen contains only certain frequencies that may be derived by a simple formula as the difference between two terms (see chapter 2). This fact should be the natural consequence of any theory describing the hydrogen atom, and it triggered Bohr to find the solution of the problem. Immediately he dedicated himself to the writing of three fundamental papers with which he designed his revolutionary theory of the atom by applying the postulates of his model to explain the formation of atomic spectra. In the first of these papers (in the second and the third he extended and completed the theory) he explained in a general way the constitution of atoms and molecules, and in great detail the hydrogen atom by expressing some postulates—later justified by the subsequently developed quantum theory—that allowed him to explain the incomprehensible facts that arose with the Rutherford model. He realized that the demand to apply laws of classical mechanics to the atom was completely unjustified. There is no reason, in fact, to think that the classical laws, developed to explain the motion of celestial bodies or of the bodies which usually surround us, should be true for bodies of dimensions one billion times smaller.

When Bohr was struggling with these problems, Planck had already established that light emission and absorption may occur only through finite quantities of energy that he called quanta, and Einstein—as we will see in the next chapters—had already given his explanation of the photoelectric effect in terms of light quanta, so Bohr thought that the principle of energy quantization was valid for any system. Therefore the mechanical energy of a system must be quantized, that is may assume only some discontinuous values, and the energy a system may exchange cannot be arbitrary but must have discontinuous values. The system may be seen as a small brick tower (figure 17) which may change its height only by a finite thickness by taking away or adding the thickness of one brick. In a similar way the energy of a system can increase or decrease, not by an arbitrary quantity but by the quantity which corresponds to the minimum quantum (that in the previous example is the brick). Of course we will notice this discontinuity if the minimum energy quantum that can be exchanged is large enough to be measured. In the majority of cases this does not happen, because the minimum quantity of energy that may be exchanged is so small that the exchanges seem to occur continuously. In systems of extremely small dimensions, this is no longer true and energy quantization becomes very important.

The electrons of the Rutherford model do not fall on the nucleus for the simple reason that they possess the minimum quantity of energy corresponding to that condition, and because this energy minimum, by

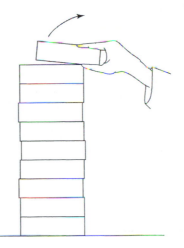

Figure 17. In quantum theory the energy of a system may have discrete values only, exactly as the height of a brick pile may change only by the thickness of a brick.

definition, cannot decrease any more, their motion must continue undisturbed for ever.

If we try to give more energy to the atom, the first quantum of this energy changes completely the state of motion of the atom and brings its electron into the so-called first excited state. In order to come back to its normal state, our atom must emit the energy quantity that it formerly received, and among the different possibilities (it could for example yield it through the collision with another atom) it may emit it in the form of a single light quantum, that according to one of the Bohr postulates has a well determined wavelength. In Bohr's theory the allowed energy states were given by a mysterious relationship that established that the angular momentum of the electron in the atom (the product of the momentum of the electron and its orbit radius) may assume only discrete values that are integer multiple of the Planck constant $h/2\pi$.

The theory yielded the formula

$$1/\lambda = 109.678(1/m^2 - 1/n^2)$$

that corresponds exactly to the Balmer formula if $m = 2$, but also predicts the other series if $m = 1, 3$, etc. Moreover, a particularly convincing argument in favour of Bohr's theory was that the coefficient 109.678 which was obtained from experimental spectroscopic evidence was exactly predicted by the theory. Light emission receives therefore a very precise explanation. It is emitted by all the atoms that, in one way or the other, have been excited, the ensuing de-excitation yielding 'light quanta' (that later will be called 'photons'). The energy emitted as light is the energy difference between the

71

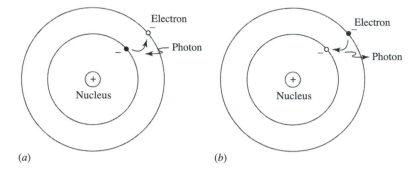

Figure 18. The photon absorption and emission processes. (a) The photon (that is absorbed and disappears) knocks an electron which sits on an inner orbit and makes it jump to an outer orbit. In (b) an electron jumps from an outer orbit to an inner one and the energy difference is emitted in the form of a photon.

excited state energy and that of the state at lower energy ('fundamental' or 'ground' state), and the photon has a frequency that is given by this energy divided by the Planck constant h. In this way Balmer's formula (difference between two terms) is automatically recovered. In fact, because the product of the frequency and the wavelength is equal to the propagation velocity of the wave, the term $1/\lambda$ that appears in the formula is proportional to frequency and therefore to the energy. According to Bohr, therefore, the electrons in an atom may exist only in certain states that Bohr represented as orbits adopted by the electron travelling around the nucleus. At variance with what was required by the classical theory, Bohr predicted that the electron when in these orbits does not irradiate energy. It emits or absorbs energy only when it passes from one orbit to the other (figure 18). The energetic states of an atom are usually represented as shown in figure 19 by horizontal lines that are at a height that depends on the energy of the level. Usually in these diagrams the lower level represents the fundamental state and the successive levels at increasing heights represent the excited levels. A transition consists of a passage from one level to another, and may be represented by vertical arrows, as shown in the figure that refers to hydrogen. Similar diagrams will be used later in our explanation of the working principles of masers and lasers.

The models developed before Bohr
As we already saw, when completely new ideas appear, their formulation often has been somehow anticipated with some of the new concepts appearing as incomplete theories or as theories that mix them with erroneous notions.

In some respects the Bohr model was anticipated.

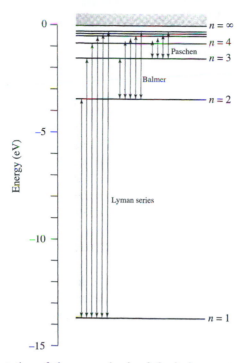

Figure 19. Representation of the energy levels of the hydrogen atom. The arrows show some transitions. On the ordinates the energy of the levels is given in electronvolts ($1\,eV = 1.6 \times 10^{-19}$ joule). The upper level with zero energy corresponds to the case in which the electron exits the atom (ionization).

In 1910 the Viennese physicist, Arthur Eric Haas (1884–1941), a doctorate student at the University of Vienna, discussed a model of the hydrogen atom in which an electron moved on the surface of a sphere of radius r, positively charged (therefore it is not the Rutherford model) having its energy quantized (this was the interesting idea).

In November 1911, John William Nicholson (1881–1955), at the time at Trinity College in Cambridge, used the recently introduced Rutherford model and recognized that the production of atomic spectra is essentially a quantum phenomenon. 'Fundamental physical laws must be found in the quantum theory of radiation, recently developed by Planck and others, according to which energy exchanges among systems of a periodic kind may only take place by some defined quantities determined by the frequencies of the system' he wrote, and discovered the form that the quantum principle must assume in its application to the Rutherford atom, that is that the angular momentum of an atom may only increase or decrease by discrete quantities. Nicholson, however, did not follow Conway's idea that only one electron at a time may emit radiation, and studied the vibrations of a great number of electrons turning around the nucleus. He assumed that an atom with a single electron could not exist and that the simplest and lightest atoms were in the order the coronium (an element discovered in the solar corona) with an atomic weight about half that of hydrogen, then hydrogen and nebulium (a hypothetical chemical element assumed to be present in some nebulae; today we know that the spectral lines attributed to it are due to oxygen and nitrogen in exceptionally high ionization states) with 2, 3 and 4 electrons, respectively. He thought moreover that helium was a compound. It was a muddle of wrong ideas: helium is an element, coronium and nebulium are not (the spectral emission attributed to coronium is actually due to ordinary metals like iron and nickel in exceptional ionization conditions) and there are no elements lighter than hydrogen.

The following year the idea of quantization of the angular momentum of the atom was examined again by the Dutch chemist Niels Bjerrum (1879–1958) and by Paulus Ehrenfest (1880–1933) who derived the correct expression in which the Planck constant h appears.

All these partial results, however, were born through trials deprived of a general vision of the problem and were mixed with completely erroneous considerations. Bohr built up his model trying to give an explanation to the many investigations that existed and succeeded even if he was not able to justify his hypotheses.

The acceptance of Bohr's hypothesis

We may ask how Bohr's theory was received. Rutherford, to whom Bohr sent the first manuscript for publication, presented it to the important English journal *Philosophical Magazine*. This suggested that he supported it, even if when Bohr sent the manuscript from Denmark he objected with his usual practical sense, 'How does the electron know to which orbit it should jump?' His argument was that in fact if the electron in the transition emits a photon that has an energy equal to the difference between the initial and final state energy, it should know its destination (final state) before the photon is emitted. To this question only Einstein gave an answer in 1916 through introducing the probability laws. In any case Rutherford suggested that Bohr shorten the manuscript, but Bohr, although he was younger and less affirmed than his teacher, refused energetically. The other European physicists expressed contrasting opinions; however, Einstein appeared enthusiastic.

In presenting his model, Bohr did not claim to give a final description of atomic systems. The break with classical physics that Bohr made with his theory was so radical that to some people his work appeared a simple numerical joke: but its ability to predict relations confirmed by experiment drew much respect. Therefore, even if it caused no immediate sensation, it was recognized little by little.

The three papers were published in the *Philosophical Magazine* between the summer and the autumn of 1913. In 1913–14, Bohr gave an informal course of lectures at Copenhagen University, while the decision to grant him a professorship was pending. During this year he made several trips to England, and in September discussed his theory at the annual meeting of the British Association for the Advancement of Science in Birmingham. The Society was established in 1831 at York, somehow in antagonism to the Royal Society. Its meetings were, however, very interesting as when in 1899 at Dover, J J Thomson announced the discovery of the electron.

At the Birmingham meeting, the announcement of new experimental evidence in support of Bohr's theory improved its reception among the originally rather sceptical British audience. However, the German mathematicians at Göttingen gave his ideas a cool reception because they criticized Bohr's application of the mathematics of classical physics to a model which defied the very basis of the classical view. In July, a trip to Germany helped Bohr to win support from that side, including the conversion of the physicist Max Born (1882–1970) who later provided a crucial link in the development of the theory by improving the matrix mechanics and contributing to the interpretation of the quantum mechanical functions. Born was awarded the physics Nobel prize in 1954 (together with Walther Bothe, a cosmic rays researcher) 'for his fundamental research in quantum mechanics, especially for his statistical interpretation of the wavefunction'. On his grave in Göttingen is engraved the fundamental equation of matrix mechanics $pq - qp = h/2i\pi$.

In the spring of 1914, Rutherford offered Bohr a readership at Manchester for the year 1914–15, later extended to 1916. In May 1916 finally he was appointed professor of theoretical physics at Copenhagen. In the autumn of 1916 his first assistant, the Dutch physicist H W Kramers (1894–1952) who remained in Copenhagen until 1926, joined him. In 1918 Oskar Klein (1894–1977) became his second assistant. In 1917 Bohr started negotiations to build a new Institute of Theoretical Physics but four years were needed before the doors of the Institut for Teoretisk Fysik were opened (8 March 1921). Through these doors one of the most astonishing series of bright scientists was to pass, as students, colleagues and guests, that the history of science has ever witnessed.

Bohr's papers on atomic structure started a great activity in many research centres, and Bohr himself contributed with further progress. A very important concept he developed to treat quantum problems, and that nobody better than him knew how to apply, was the 'Correspondence Principle' that relates predictions of classical theory with quantum theory. As Planck's quantum formula for long wavelengths is well approximated by the classical Rayleigh formula, so Bohr argued that the classical mechanical frequency of rotation of the electron along its orbit, for very large orbits, should be well approximated by the formulae given by the classical laws.

This allowed him to find rules—called selection rules—that established that not all transitions could take place, and to discover between which orbits transitions were allowed, therefore establishing the first criteria to predict which frequencies could be emitted (among the many corresponding to the different energy jumps). The rules also facilitated predictions of the light intensity corresponding to each possible transition to be made.

In June 1922 he gave a series of lectures at Göttingen where he met Wolfang Pauli (1900–1958) and Werner Heisenberg (1901–1976) who were with him in Copenhagen for the next five years, and contributed to the new revolution made by quantum mechanics.

In December 1922 he was awarded the physics Nobel prize 'for his services in the investigation of the structure of atoms and of the radiation emanating from them'.

In the ten successive years he was very busy with the direction of the ever expanding Institute which became more and more the place for all progress in atomic physics.

In their reminiscences, physicists who worked at Bohr's Institute between the World Wars, emphasize the unique 'Copenhagen spirit' of scientific investigation. They remember this spirit as, first, an unlimited freedom to pursue whatever problems in theoretical physics that they considered the most urgent. The second aspect of the Copenhagen spirit was that this pursuit took the form of intense discussions between Bohr, the acknowledged master, and the most promising, yet young and unestablished physics students visiting his Institute from several countries. Needing to discuss, to develop his ideas, Bohr encouraged some visitors to become his 'helpers', that is to take part in his own thinking process. The Copenhagen spirit thus comprised a complete freedom of research pursued with a division of scientific labour between Bohr and the cream of international theoretical physics students.

In fact while the new quantum mechanics was appearing, Bohr applauded the dramatic progress being made but also pointed out the inconsistencies between classical and quantum theories.

In collaboration with Kramers and John Slater (1900–1976), he published in 1924 what was to be the last attempt to describe an atomic system along the quasi-classical lines he had employed earlier. In that work the authors suggested that in individual atomic interactions, energy was not conserved. Although this suggestion was quickly refuted by experiment, the revolutionary character of the proposal shows how desperate Bohr considered the situation at this time.

The same year Kramers succeeded in formulating a mathematical theory to explain the dispersion of light by atoms. Working from this base, Heisenberg developed a purely abstract mathematical representation of quantum mechanical systems.

Throughout 1925–26, Heisenberg refined and extended his theory with the help of Max Born and Pasqual Jordan (1902–1980), producing what today is known as 'matrix mechanics'. That spring, the Austrian physicist Erwin Schrödinger (1887–1961), working quite independently, put forward a 'wave mechanics' representation of quantum systems later demonstrated to be mathematically equivalent to Heisenberg's matrix mechanics. These two different approaches convinced Bohr that mathematically the theories were on the right track, but at the same time they raised his concern about the physical interpretation of the mathematical formalism. Bohr was more worried than anybody about the inconsistencies of quantum theory.

In 1926–27, Heisenberg returned to Copenhagen to discuss the problems that worried Bohr. Also Schrödinger visited the Institute that autumn, and in the following discussions Bohr convinced himself to accept the application of the wave–particle dualism of light phenomena, already strengthened, as we will see, by Einstein, in the interpretation of atomic systems. Working in Copenhagen in February 1927, Heisenberg formulated the 'Uncertainty Principle' according to which it is not possible to measure with a precision as great as one wishes both the position and velocity of a particle. At the same time, while skiing in Norway, Bohr began to put forward the basis for the 'complementary principle'.

The basic principle of this concept is very simple even if very strange. It says that we may ask nature some question, for example what is the position of an electron, or a complementary question, that is how great is its momentum (essentially the velocity), but nature is so constructed that asking one question automatically excludes the possibility of asking simultaneously the complementary question. Quantum mechanics was founded upon the different theories of Heisenberg and Schrödinger and had identified the existence of the wave–particle dualism of light and matter. Bohr realized that our models of matter and light are based upon their behaviour in various experiments made in our laboratories. In some experiments, such as the photoelectric effect that we will discuss shortly, light behaves as if it consists of particles. In other experiments, such as those concerning interference, light behaves as if it consists of waves. Similarly, in experiments such as J J Thomson's cathode ray studies, electrons behave as if they are particles; in other experiments, such as diffraction studies, electrons behave as if they are waves. But light and electrons never behave simultaneously as if they consist of both particles and waves. In each specific experiment they behave either as particles or as waves, but never as both.

This suggested to Bohr that the particle and wave description of light and of matter are both necessary even though they are logically incompatible with each other. They must be regarded as being complementary to each other. Each experiment selects one or the other description as the proper one.

Complementarity had a place in practically all Bohr's discussions. When he was knighted with the Order of the Elephant, he was requested to propose a heraldic phrase and he chose '*Contraria sunt complementa*'.

Werner Heisenberg recollected that at the time he wrote his work on the Uncertainty Principle, once he sailed with Bohr and Niels Bjerrum, and he was explaining to Bjerrum the content of his paper. After listening to him, Bjerrum addressed Bohr saying 'But Niels, this is what you have been saying to me since we were boys!'.

Complementarity was first presented at the International Physics Conference in Como in 1927 for the centenary of the death of Alessandro Volta. It was a very important meeting in which quantum mechanics was seriously discussed for the first time en masse. Many of the most outstanding physicists were present; only Einstein had not wanted to enter fascist Italy. Although Bohr had worked all summer on his manuscript, the paper he delivered was very far from its finished form. Most of the audience were unimpressed, finding Bohr's argument far too philosophical and including nothing new in physics. Pauli recognized the significance of the new ideas, and worked with Bohr in Como, after the conference, to produce a more refined manuscript. After further work the definitive version was completed by Easter 1928. Meanwhile, in October 1927 Bohr had an opportunity to present complementarity at the Solvay Congress in Brussels. All the great men of European physics were present, Einstein included. Einstein's reaction was strongly negative and produced a series of discussions that lasted for years.

With the discovery of the neutron and the development of a theory of the nucleus, Bohr had shifted his interest to the application of quantum theory to nuclear phenomena. In 1935 he formulated a theory of nuclear reactions and successively in 1939 the first theory of fission phenomena, together with J A Wheeler (1911–).

In 1943 the Nazis occupied Bohr's Institute, and in September of the same year Bohr escaped in an extraordinary way.

At the beginning of 1943 Bohr had received a visit from Captain Volmer Gyth, an official of the information services of the Danish General Staff and at the same time a member of the Danish resistance movement. Gyth asked Bohr if he was willing to receive a message that would come from England. Some time later the message arrived: it was written on a microfilm hidden in a cavity the size of a pin-head made in a key. Gyth, who Bohr had asked to be present, took the microfilm and read it with a microscope. It was a message from the English physicist J Chadwick (1891–1974, physics Nobel prize in 1935 'for the discovery of the neutron') who invited Bohr to transfer to England. At the time Bohr replied that he could not abandon his collaborators at the Institute. The letter was reduced to the dimensions of two by three millimetres, folded in metallic foil and consigned to a courier. The man went to a dentist who inserted the message in a hollow tooth, then covered it with a filling. However, the situation worsened quickly and in mid-September Bohr was informed that the Germans were planning to put him in prison. He contacted the Danish underground movement which organized his escape, and on 29 September Bohr and his wife left their residence and walked to a farmhouse where they arrived in the early evening. From there together with others among whom were his brother Harald and his son, at ten o'clock at night, crawling on all fours, they were directed to the shore where they boarded a small fishing boat that brought them onto the Oresund. Then they transferred to a large trawler, which in the first hours of 30 September disembarked at Limhamn harbour from where they were transported to Malmo. Here they were lodged in the cells of a police station. In the early afternoon of the same day, Bohr reached Stockholm by train where he was met by, among others, Captain Gyth, who transmitted immediately to England confirmation of his escape.

The transfer of Bohr to England was then organized. Gyth accompanied Bohr to Professor Klein in Stockholm first by taking a taxi that brought them to a building used by the Swedish intelligence service, then by walking over the roofs they reached another building from which they took another taxi. Bohr remained well-protected in Stockholm for several days, meeting various personalities and discussing the Jewish problems and eventually left on 4 October.

After his departure, Gyth and Bohr's hosts broke open a bottle of Champagne to celebrate the success of the venture, but somehow after midnight there came a call from Bohr who was again back in Stockholm. The plane had had engine problems and had been forced to return. Gyth had remained alone, all the security personnel had been dismissed and he therefore decided to guard Bohr's bedroom by himself, armed with an old revolver. Finally the following night, Bohr took off again. Notwithstanding the secrecy of the whole operation, the *New York Times* on 9 October announced that Bohr had arrived in London from Sweden bearing the plans for a new invention involving atomic explosions. On 6 December he arrived in the United States together with his son Aage—to be awarded, later on in 1975, with the Nobel prize in physics together with B. R. Mottelson (1926–) and Leo James Rainwater (1917–1986) 'for the discovery of the connection between collective motion and particle motion in atomic nuclei and the development of the theory of the structure of the atomic nucleus based on this connection'—and took part, in a peripheral way, in the development of the atomic bomb to which he had made a vital contribution with his nuclear fission theory.

He died on 18 November 1962. In 1965 the Theoretical Physics Institute was given his name.

Bohr had perhaps more than anyone else a fundamental influence on the development of contemporary quantum theory, and was the originator of the way in which quantum results are today interpreted, referred to as the 'Copenhagen school interpretation'. He first realized that his theory was only the first step toward the solution of the problem, having no firm logical basis and, notwithstanding its success, he insisted in putting it on solid foundations, and did not mask the contradictions that it contained. When he died, in 1962, quantum theory had been completely developed by

Louis de Broglie (1892–1987), Nobel prizewinner in physics in 1929 'for his discovery of the wave nature of electrons', Werner Heisenberg, Nobel prizewinner in physics in 1932 'for the creation of quantum mechanics, the application of which has, *inter alia*, led to the discovery of the allotropic forms of hydrogen', Erwin Schrödinger and Paul A M Dirac (1902–1984), who both were awarded the Nobel prize in physics in 1933 'for the discovery of new productive forms of atomic theory', and many others, and had completely explained the nature of the atom, the process of emission and absorption of light providing the logical basis to his bright intuitions.

The life and personality of Bohr have been described in a great number of biographies. He dedicated a great effort to the composition of his papers that had always a long and laborious gestation period. He was moreover repelled by the material operation of writing. His first papers, among which is his dissertation, were dictated to his mother, then to his wife and finally to a long series of collaborators starting from Kramers. At the time he was preparing his atom theory, week after week, Bohr did not publish. Rutherford pushed. Bohr protested: 'Nobody will trust me unless I can explain every atom and molecule'. Rutherford immediately replied: 'Bohr, explain the hydrogen atom, explain helium and everyone will believe the rest'. His habit of dictating gave rise once to an amusing fact reported by the theoretical physicist Abraham Pais (1918–2000), the author of a successful and exhaustive biography. Bohr was reordering a speech he had to make on the occasion of the third centenary of the birth of Newton:

'He stood in front of the blackboard (wherever he dwelt, a blackboard was never far) and wrote down some general themes to be discussed. One of them had to do with the harmony of something or other. So Bohr wrote down the word harmony. It looked more or less like this:

'However, as the discussion progressed, Bohr became dissatisfied with the use of harmony. He walked around restlessly. Then he stopped and his face lit up. 'Now I've got it. We must change harmony to uniformity'. So he picked up the chalk again, stood there for a moment at what he had written before, and then made a single change:

'With one triumphant bang of the chalk on the blackboard'.[1]

With Einstein, Bohr had very friendly relations, but since he announced the Complementarity Principle the two men had continuous discussions on the meaning of quantum mechanics. Once, during a visit to the Institute for Advanced Study in Princeton, of which he was a permanent member, Bohr wanted to note something down on the argument and as usual he needed a secretary. He called Pais who was also in the Institute and asked him to sit down:

> 'And soon started to pace furiously around the oblong table in the center of the room [Pais recollected]. He then asked me if I could note down a few sentences as they emerged during his pacing. It should be explained that, at such sessions, Bohr never had a full sentence ready. He would often dwell on one word, coax it, implore it, to find the continuation. This could go on for several minutes. At that moment the word was "Einstein". There was Bohr, almost running around the table and repeating "Einstein...Einstein..." After a little while he walked to the window, gazed out, repeating every now and then: "Einstein...Einstein...".
>
> 'At that moment the door opened very softly and Einstein tiptoed in. He indicated me with a finger on his lips to be very quiet, an urchin smile on his face. Always on tiptoe he made a beeline for Bohr's tobacco pot, which stood on the table at which I was sitting. Meanwhile Bohr, unaware, was standing at the window, muttering "Einstein...Einstein...".'

Then Bohr turned around with a firm 'Einstein' and the two men were face to face. Bohr remained silent, and Einstein explained that his doctor had forbidden him to buy tobacco, but not to steal it, and this was precisely what he was setting to do. Needless to say the three men burst out with laughter.[2]

Now it is time to speak about Einstein and his contributions to the theory of light.

[1] A. Pais, *Niels Bohr's Times*, Oxford 1991, page 10.
[2] A. Pais, ibidem page 13.

CHAPTER 5

EINSTEIN

Albert Einstein is universally known for the relativity theory he elaborated and developed between 1905 (the year of the formulation of Special Relativity) and 1915 (General Relativity theory). Only specialists know the fundamental contributions he provided, almost in the same period (1905–1916), on the nature of light: contributions that bore relevance to the invention of the maser, and later of the laser.

Novalis, the science enthusiast and romantic German poet of the 18th century, said: 'Theories are like fishing-nets: only he who casts will catch'. If we extend this metaphor we may say that no fisherman in the 20th century had greater success than Einstein. In 1905, his *annus mirabilis* (comparable in the history of science maybe only to the memorable year of 1666 in which Isaac Newton conceived most of the ideas that ruled science for more than two hundred years) Albert Einstein published, in the same volume of the German scientific journal *Annalen der Physik*, three papers, each one of which, besides containing important scientific results, put forward the basis for new and large fields of fundamental research, as we will describe in the following.

The young Einstein

Who was this man, Albert Einstein, who when a third class expert technician of the Swiss Patent Office in Bern, at the age of 26, invented in his free time new methods of statistical mechanics, introduced light quanta, produced a proof of the existence of atoms, and solved the problem of correctly formulating the electrodynamics of moving bodies, a problem that had been confronted with no success by the most influential researchers of the time such as Hendrik Anton Lorentz and Henri Poincaré (1854–1912), so constructing a new theory of space and time?

German by nationality, Jewish by origin, and challenger by vocation, Einstein reacted ambivalently against these three natural gifts. He threw his German nationality overboard at the age of 16; 20 years later, after becoming Swiss, he settled in Berlin where he remained throughout the First Word War; after the defeat suffered by Germany in 1918, he resumed his fight for German civil rights, again to give up a second time when Hitler rose to power. The fact he professed Zionism was the confirmation he felt Jewish, but that was an adhesion to which he failed more than once.

He was born on 14 March 1879 in Ulm, an ancient German city which in 1805 witnessed the rout of Austrians defeated by Napoleon, in the same year in which were born Max von Laue (1879–1960)—the discoverer of x-ray diffraction—Nobel prize in physics 'for his discovery of the diffraction of x-rays by crystals', and the chemist Otto Hahn (1879–1968, the 1944 Nobel prize winner in chemistry 'for his discovery of the fission of heavy nuclei') who discovered radiothorium and protactinium and who together with Lise Maitner (1878–1968) and Fritz Strassmann (1902–1980) discovered nuclear fission, and in which James Clerk Maxwell, the founder of the modern theory of electromagnetism, died.

Einstein's family, who came from Buchau, a small quiet town along the route to Lake Constanz, were Jewish but no longer followed Jewish religious rites and laws. In 1880 they settled in Munich, where his father, the engineer Hermann, opened a small electro-chemical workshop in partnership with his brother Jacob. The following year, Albert's only sister was born, Maja, to whom Albert was closely attached.

The young Albert has been described as a reserved, pensive, dreaming boy, who began to speak at a late age, did not like physical activities and was unwilling to play with other boys. When he was four or five years old, something that struck him deeply occurred: his father showed him a pocket compass that, as if it was constantly guided by an invisible and mysterious power, pointed always in the same direction, whichever way the case was turned.

A short time later when he was five or six, he began to learn the violin with a great passion, even if his delight was never quite matched by performance. He attended a Catholic elementary school, since it was the most convenient, and found himself a Jew among Christians; among Jews he was, like the members of his family, an outsider. Later, at the age of ten he was transferred to the Luitpold Gymnasium, but was intolerant of the firm discipline, dictatorial spirit, and the loss of freedom in the gymnasium where he was forced to study Latin and Greek grammar which interfered with his studies of mathematics and physics, an interest that he derived from his uncle, the engineer Jacob. 'Algebra is a merry science', Uncle Jacob used to say. 'We go hunting for a little animal whose name we don't know, so we call it x. When we bag our game we pounce on it and give it its own name.'

At that time Einstein had read a book on Euclidean geometry and a series of popular books on science and had familiarized himself with the principles of differential and integral calculus. However, he was unhappy and depressed, an ill-fitting outsider, accused by his teachers of having a disruptive influence on his classmates; but even if his childhood was unexceptional, his scholarly life, contrary to what is written in some biographies, was brilliant enough.

At the Luitpold Gymnasium, a teacher told him he would have been happier if he was not in his class, and when Einstein remarked he was doing nothing he replied: 'Yes, that is true. But you sit there in the back row and smile, and that violates the feeling of respect which a teacher needs from his class'. At this time, in the young boy grew a spirit of hostility with respect to officialdom and to imperial Germany that never left him.

The initially prosperous workshop of his father collapsed, and solicited by their agent in Italy, the brothers Hermann and Jacob Einstein settled with their families first in Milan in 1894 and the following year in Pavia, where they opened a new workshop. Albert was left in Munich to complete his studies but at the beginning of spring 1895, with a certificate from the family doctor that declared a nervous breakdown, he obtained permission to leave the Luitpold Gymnasium he hated and rejoin his parents in Italy who knew nothing of his initiative.

He promised his parents that he had prepared alone for the entrance examination for the Zurich Polytechnic and informed them that he intended to renounce his German citizenship, as he indeed did later when attending the Polytechnic. It is also possible that he left Germany so precipitously to escape serving in the German army, being a firm antimilitarist.

Zurich Polytechnic

Zurich University was established in 1833 in response to the request of the local population for an institute of higher education. At this time the only Swiss University was in Basel. In 1855, to satisfy the need for technical education, the Eidgenossische Technische Hochschule (Federal Institute of Technology; the Polytechnic, usually indicated with the abbreviation ETH) was established, and Rudolf Clausius (1822–1888) was appointed professor of physics (1857). Clausius was one of the great physicists of the 19th century. He established the second law of thermodynamics and defined the concept of entropy; also he provided important contributions to the kinetic theory of gases.

In 1878, Alfred Kleiner (1849–1916) was appointed professor of experimental physics. His major achievement, as he himself often used to admit, was his approval in 1905 of Albert Einstein's dissertation, and the lobbying in 1909 for Einstein to be given the chair of theoretical physics, the first one in that University.

In October 1895, Einstein, as a result of lacking the standard school-leaving certificate, was refused entry by the Polytechnic and was not even allowed through the entrance examination stage, notwithstanding his excellence in mathematics and physics. To obtain the necessary school certificate he enrolled at the cantonal school at Aarau, in the German-speaking region of Switzerland, where he was much happier than in the Luitpold Gymnasium, and the following year he entered the Polytechnic as a student of mathematics and physics. In 1901 he adopted Swiss citizenship.

During the year he passed in Aarau, he posed himself a problem: if one runs after a light wave with a velocity equal to the light velocity, then one would encounter a time-independent wavefield. However, such a thing does not exist! This was the first thought experiment he considered and the paradox revealed by it, after ten years of reflection, led him to the theory of Special Relativity.

A short essay written when he was studying at Aarau gives us an idea of his projects:

> My projects for the future
> A happy man is too content with the present to think much about the future. Young people, on the other hand, like to occupy themselves with bold plans. Furthermore, it is natural for a serious young man to gain as precise an idea as possible about his desired aims.
> If I were to have the good fortune to pass my examinations, I would go to [the ETH] Zurich. I would stay there for four years in order to study mathematics and physics. I imagine myself becoming a teacher in those branches of the natural sciences, choosing the theoretical part of them.
> Here are the reasons which led me to this plan. Above all, it is [my] disposition for abstract and mathematical thought, [my] lack of imagination and practical ability. My desires have also inspired in me the same resolve. That is quite natural; one always likes to do the things for which one has ability. Then there is also a certain independence in the scientific profession which I like a great deal.

Among his fellow-students at Polytechnic he met Mileva Maric (1875–1948) a dark-haired Serbian girl four years older than him who in 1903 became his first wife and later the mother of his three children, and the Swiss Marcel Grossmann (1878–1936)—a classmate—who 18 years later was his mathematical collaborator in the writing of the theory of general relativity. Among his teachers were the famous mathematician Hermann Minkowski (1864–1909) who, in 1907, invented the concept of space–time, contributing in an essential manner to the development of the theory of relativity.

At Polytechnic—where he became a friend of Michele Angelo Besso (1873–1955) a young engineer from Trieste who was his colleague from 1904 at the Patent Office, remaining a close friend and correspondent for all his life—Einstein worked for most of the time in the physical laboratory, fascinated by the direct contact with experiment. However, his teacher, the professor Heinrich Friedrich Weber (1843–1912) whose lectures in physics were disliked by Einstein, was not particularly enthusiastic and at one time told him 'You are a smart boy, Einstein, a very smart boy. But you have one great fault: you do not let yourself be told anything'.

During the last semester, as a result of hearing the Hermann Minkowski lectures on capillarity, Einstein was stimulated to work on this problem. Capillarity is a special form of energy connected with the shape and position of the surface of fluids. So, for example, it determines the curved shape that may be observed in the level of a liquid in a thin tube (capillary). In the 19th century many scientists, amongst whom were Thomas Young, Pierre Simon de Laplace (1749–1827), Carl Friedrich Gauss (1777–1855), James Clerk Maxwell, Josiah Willard Gibbs, Johannes Diderik van der Waals (1837–1923)—Nobel prize in physics in 1910 'for his work on the equation of state for gases and liquids'—and Henri Poincaré had concerned themselves with this problem. Laplace attributed the origin of capillarity to the existence of cohesive forces exerted by the fluid molecules on each other. As a consequence of an experimental determination of the capillarity of a fluid, information about the intermolecular forces can be derived. It was this possibility that interested Einstein in his first studies in 1901 and continued to interest him even later, as we will see.

The Patent Office

After taking his degree in 1900, Einstein failed to secure a position at the Polytechnic, where he had not been able to adapt to the study of topics that did not interest him, was not particularly liked by his teachers, and after a fruitless search for a job, in 1902 with the help of his friend Marcel Grossmann he found a post in the Patent Office in Bern. In the Patent Office he felt rather content: he was taking his work seriously and often found it interesting. Moreover there was always enough time and energy left for his own physics. So he started to write physical papers that he sent to the *Annalen der Physik*, at the time directed by Wien. Among these, in 1903–04 he published some useful papers on the fundamentals of statistical mechanics that, unknown to him, however, were anticipated by Gibbs, and prepared himself for the doctorate exams, fulfilling the requirements in 1905. He continued his research in theoretical physics, writing in this same year a paper on light quanta which was to earn him the Nobel prize, the first paper on the theory of relativity, his dissertation dedicated 'to my friend Marcel Grossmann' in which he described a new theoretical method to determine the radii of molecules and the number of molecules that may reside in a given volume (the Avogadro number) and finally presenting at the same time a study on the motion of particles in suspension in a solution (Brownian motion). This last study may be considered a by-product of his thesis that would be published the following year, in 1906, also in *Annalen der Physik*.

Apart from the fundamental character of some of the results reported in his thesis, there is another reason it attracts uncommon interest. It has relevance to more practical applications than any other paper Einstein wrote. Referring to the properties of particles in a suspension, it contains results that are applied to the motion of sand particles in mixtures of concrete (important in the construction industry), of micelles of casein in milk (food industry), of aerosol particles in clouds (ecology) etc.

Einstein remained in his post in Bern until the end of 1909, when he received his first full time academic position as an associate professor at Zurich University. At this time his scientific merits were already largely recognized. Besides the results obtained on light quanta, Brownian motion and relativity, Einstein two years later published the first quantum theory of specific heat in solids. The theory of heat as energy of motion, both of particles colliding in the gas and of internal vibrations of solids, had obtained considerable success, but before 1900 it had encountered serious difficulties. Statistical mechanics, in fact, allows the calculation of how much heat must be given to a body to increase its temperature by one degree (the so-called specific heat). In the case of solid bodies this quantity theoretically was expected to be about the same for all bodies and independent from temperature. Experiment contradicted this result, showing that the specific heat increases with increasing temperature, reaching the value predicted by statistical mechanics only at high temperatures (the Dulong–Petit law). In 1907 Einstein concluded that if Planck's idea was to be taken seriously, it should be valid for all kinds of vibrations, and by applying this concept to atomic vibrations, he derived the correct behaviour of the specific heat with temperature. Again in 1907 he was asked by Johannes Stark (1874–1957), who was the director of the *Jahrbuch der Radioaktivitat und Elektronik*, to write a review paper on relativity. During the preparation of this important paper, Einstein remembers that when sitting in the Patent Office in Bern he thought 'If a person falls freely, he does not feel his own weight'. For an observer falling freely from the roof of a house there exists— at least in his immediate surroundings—no gravitational field. Indeed, if the observer drops a body then this remains relative to him in a state of rest or of uniform motion, independent of its nature. The observer therefore has the right to interpret his state as 'at rest'. Thanks to this intuition, the peculiar experimental law that in a gravitational field all bodies fall with the same acceleration, already found by Galileo, acquired at once a deep physical meaning. The observer has no element that allows him to establish that he is in free fall in a gravitational field. On the grounds of this thought, Einstein was pushed towards a theory of gravitation. He therefore reached the conviction that a satisfactory theory of gravitation should incorporate in a fundamental and natural way the equality between the inertial and gravitational mass and the fact, already observed by Galileo, that all bodies fall with the same acceleration. Gravity and inertia are essentially the same thing, Einstein decided, and therefore a satisfactory theory of gravity required a generalization of the space–time structure of his theory of relativity, because if gravity is taken into account the concept of a finite and strictly inertial rest frame is no longer adequate.

His academic career

Sometime after December 1907 Einstein started his academic career. The first step was a request, as usual at the time, to obtain permission to teach

(Privatdozent) in the University, which if granted allowed only nominal sums to be received from the students. The application was rejected by Bern University because of a formality. Einstein had omitted to enclose in his documentation (the doctorate thesis and 17 published papers) the Habilitation thesis (a scientific paper not yet published). At the beginning of 1908 he submitted the necessary paperwork and received the title.

He was still working, however, at the Patent Office and therefore was obliged to hold lectures at strange times. In 1908 he delivered lectures on Saturday and Tuesday mornings from 7 to 8 a.m. to three students, one of whom was Besso, who worked with him in the Patent Office. In 1908–09 he delivered a second and last course every Wednesday night from 6 to 7 p.m. to four people.

Professor Einstein

Finally in 1909 Einstein was awarded the first chair as an associate professor of theoretical physics at Zurich University. It was a new post: there had been no further theoretical physics professorial appointments since Clausius left the University in 1867. Einstein was presented by Professor Alfred Kleiner who spoke very well of him and in the minutes of the Faculty meeting one may read:

'These expressions of our colleague Kleiner, based on several years of personal contact, were all the more valuable for the committee as well as for the faculty as a whole since Herr Dr. Einstein is an Israelite and since precisely to the Israelites among scholars are ascribed (in numerous cases not entirely without cause) all kinds of unpleasant peculiarities of character, such as intrusiveness, impudence, and a shopkeeper's mentality'.

On 6 July 1909, Einstein submitted his resignation to the Patent Office and moved to the University. Professor Einstein appeared in class in casual dress, often wearing trousers which were too short and bearing a small piece of paper the size of a visiting card on which he had written his lecture notes. Between 1907 and 1911 Einstein had lost interest in gravitation theory. He was instead totally absorbed by quantum theory. In 1908 he wrote to his collaborator J J Laub (1882–1962):

'I am incessantly busy with the question of the constitution of radiation.... This quantum problem is so uncommonly important and difficult that it should be the concern of everybody.'

And the following year:

'I have not yet found a solution of the light-quantum question. All the same I will try to see if I cannot work out this favourite problem of mine.'

However, later, he abandoned temporarily his efforts on light theory to come back to the theory of gravity. In 1910 Einstein accepted a chair at the German University of Prague where he transferred in March 1911. Now he was trying to generalize the theory of special relativity to encompass gravitation. The theory of gravitation was his major interest until 1916. While most physicists had already accepted special relativity theory as a solid part of the foundation of physics, Einstein was busy finding its limits of validity and a mathematical representation more global and more representative of many physical phenomena.

In Prague, in 1911, he arrived at the prediction that light waves are curved by gravitational fields, but it was necessary to wait until 1914 before an expedition was ready to make the necessary observations to check the prediction during a solar eclipse. The First World War impeded these observations, however, and the first measurements could be performed only in 1919.

In 1911 Einstein was also busy preparing an important lecture on quantum physics at the First Solvay Congress (30 October–3 November 1911). After eighteen months in Prague, Einstein returned to Zurich at the end of 1912, this time as full professor at the Polytechnic where he had studied a dozen years before. In Zurich, in collaboration with Marcel Grossmann, who had become mathematics professor, Einstein published in 1913 a preliminary version of a new theory of gravitation.

At the end of 1913, under the initiative of the German physicists Max Planck and Walther H Nernst (1864–1941)—Nobel prize in chemistry in 1920 'in recognition of his work in thermochemistry'—he was offered a well rewarded position as a member of the Royal Prussian Academy of Sciences in Berlin, a professorship at the University of Berlin without obligation to teach, and the directorship of the newly founded Kaiser-Wilhelm-Institut für Physik. His task was to organize research. He was not forced to teach, but could do so if he was willing. Einstein had always considered formal teaching a bore, and was attracted by the lively scientific atmosphere of Berlin. Therefore he accepted the offer.

During an exchange with Max Planck on the occasion of the first visit he and Walther Nernst made in Zurich to offer Einstein the new position, on Planck's request, Einstein described the status of his work on general relativity and Planck, who had been the first to recognize his genius, told him: 'As an older friend I must advise you against it for in first place you will not succeed; and even if you succeed, no one will believe you'.

When Planck and Nernst had left, Einstein commented towards his assistant Otto Stern: 'The two of them were like men looking for a rare postage stamp'.

A little after his arrival in Berlin, Einstein separated from his wife Mileva; he was 34 and was a star of the first magnitude in the heavens of science.

In Berlin, notwithstanding his many contacts with colleagues, in particular with Max Planck, Max von Laue, Walther Nernst, and later Erwin Schrödinger and many others, he felt isolated and a stranger. He did not lecture but participated actively in the discussions that followed scientific seminars. As a pacifist and opponent of nationalism, he felt even more isolated during the First World War. He concentrated completely on the theory of gravitation and, after a strenuous effort, succeeded at the end of 1915 in formulating a coherent theory that still today is considered as the most admirable part of classical physics. This theory has stood up to all the experimental verifications performed until now.

In 1915 he became interested also in an experiment carried out together with the Dutch physicist Wander Johannes Haas (1878–1960), the son-in-law of Lorentz, on a torque induced in a suspended cylinder (for example of iron) as a consequence of its being abruptly magnetized; today this is known as the Einstein–de Haas effect.

Einstein's productivity was not affected by the deep troubles of the war and indeed those years rank among the most productive and creative of his career. During this period he published a book and about 50 papers. In 1916 Einstein wrote ten scientific papers, among which was his most important synthesis of general relativity, the discussion of the theory of light emission with the introduction of spontaneous and induced emission, the first paper on gravity waves, and others we will discuss later. He also finished his first semi-popular book on relativity.

In this year he came back once more to blackbody radiation and made further progress. In November 1916 he wrote to Besso: 'A splendid light has fallen on me about the absorption and emission of radiation'. His reasoning is divided into three papers, two of which appeared in 1916 and the third one early in 1917. In these papers, which we may consider as his most important contribution to quantum theory, Einstein proposed a statistical theory of the interaction between atoms and photons, gave a new demonstration of Planck's radiation theory and introduced the concept of 'stimulated emission', providing the basis for the discovery of masers and lasers, of which we will speak in the next chapter. In that same year he founded modern cosmology, the science of the large scale structure of the universe, by constructing the first mathematically correct model of a universe containing matter uniformly distributed and gravitating.

Einstein's private life

After the strenuous work of previous years, in 1917 Einstein fell seriously ill. His cousin Elsa Einstein (whose marriage with a merchant named Lowenthal ended in divorce) took care of him and in June 1919 Albert and Elsa married. Elsa, who died in 1936, was happy to take care of Einstein; she gloried in his fame. Einstein the gypsy had found a home and in some ways this did him much good. He very much liked being looked after and enjoyed receiving people at his apartment: scientists, artists, diplomats and other personal friends. In other ways, however, this life was too much for him. A friend who visited him wrote:

> 'He, who had always had something of the bohemian in him, began to lead a middle-class life ... in a household such as was typical of a well-to-do Berlin family ... in the midst of beautiful furniture, carpets, and pictures. ... When one entered ... one found Einstein still remaining a "foreigner" in such a surrounding—a bohemian guest in a middle-class home.'

What we know at present of his private life has been coloured in the last years with the availability of his private letters, and the death of some who opposed their publication. Accepting that these private episodes in no way limit our judgement on his scientific production, we know that when a student Einstein fell in love with Mary Winteler, the young daughter of his Greek and History professor and his landlord at the time of the Aarau school, writing her burning love letters but giving up rapidly when she eventually returned them. With Mileva, a tumultuous relationship existed, with a daughter born in 1902 before marriage and then separated from her mother (probably given up for adoption) and Einstein's family strongly opposed to Albert's choice. The marriage ended with the appearance of his cousin Elsa of whom, after the marriage, he tired, diverting his attention to other women. Looking for a scoop, some scholar has also advanced the hypothesis that Mileva had some part in the creation of the theory of special relativity, but these deductions are not supported by any correspondence or in what we know of the scientific life of Mileva that was uneventful. In 1939, when divorcing, Einstein left the custody of his two sons to his wife and continued to support all three. He decided moreover to give to Mileva the Nobel prize money that both were sure one day he would be awarded.

Relativity theory

The theory of relativity, that constitutes a revolution in our concepts of space and time and from which very important consequences result, remained unknown, except to a few specialists until 1918, at the end of the First World War. Its dissemination then followed, attracting great attention: in fact it presented the world through a new way of thinking, a new philosophy.

It came at a time at which everyone was bored with the war: the victorious and the defeated. People wanted something new. Relativity offered exactly what was needed and it became the central argument of conversations. It allowed people to forget for a while the horrors of war and the problems that had resulted.

A fantastic number of papers were written in newspapers and journals concerning relativity. Never before or since then had a specific idea attracted such great interest. Most of what was said or written referred to general philosophical ideas and didn't involve serious scientific discussion. Little precise information was at hand; however, many people were happy to expound their ideas.

In the UK only one person, the astronomer and mathematician Sir Arthur Eddington (1882–1944) really understood what relativity was, and became the authoritative leader in this field in his country. He was very interested in the astronomical consequences of the theory, and in the possibility of verifying the theory through astronomical observations. Three possible verifications of the theory were proposed based on the predictions Einstein had made in his paper of 1915. The first one regarded the motion of the planet Mercury. The perihelion (the point of the trajectory of the planet that is nearest to the Sun) of Mercury had been observed to move, advancing about 43 arc seconds more than it should according to Newton's theory, a fact that had left astronomers perplexed for a long time. Einstein's new theory predicted exactly this effect, and Eddington's measurements confirmed these predictions. This was a great success for the theory but Einstein was unmoved when he heard of Eddington's confirmation because he was convinced his theory was right in any case.

The second confirmation regarded the bending of light that passes close to the Sun. Einstein's theory of gravitation requires that light travelling near the Sun should be deflected. Also Newton's theory predicts a deflection, but for a quantity that is one half that predicted by Einstein (figure 20). Therefore by observing the stars visible beside the solar disc, whose light has travelled near to the Sun before reaching Earth, it is possible to verify the theory. We may observe the stars near to the solar disc only at the moment of a total eclipse, when the Sun's light is blocked by the Moon. A suitable eclipse occurred in 1919 and Eddington mounted two expeditions to observe it: one to Sobral in Brasil, led by Andrew C Crommelin (1865–1939) from the Greenwich Observatory, and one to Principe Island, off the coast of Spanish Guinea, led by himself. Both expeditions obtained results

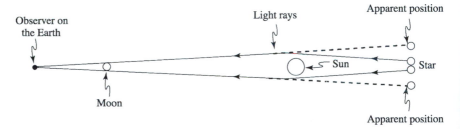

Figure 20. During a Sun eclipse, one may observe the light coming from two lateral stars. Because the Sun's gravitational field bends the light, the stars observed from the Earth on the prolongation of the rays seem to be farther apart than they really are.

that confirmed Einstein's theory. In London, on 6 November, at a joint meeting of the Royal Society and the Royal Astronomical Society, the president of the Royal Society, the Nobel prize-winner J J Thomson, after hearing of Eddington's results, celebrated Einstein's work as 'one of the highest achievements of human thought'.

However, the precision of this confirmation was not very high due to the difficulties in making these observations. More recently this effect has become verifiable using microwaves rather than light. Stellar-type objects that emit strongly in the radio frequency range (quasi-stellar radio sources, or quasars) have in fact been discovered and when one of them is behind the Sun we may observe if the radio waves that travel near to the Sun are deflected. This can be done without waiting for a total eclipse because the Sun is itself only a feeble source of radio waves. The result, once suitable corrections have been made due to the presence of other phenomena, is that Einstein's theory is confirmed with a much higher precision than with light waves.

The deflection of light by gravitation has been proven even more spectacularly in recent years. The mass of a galaxy may in fact act as a lens and focus light that comes from a distant source behind it (figure 21). If the source, the galaxy which acts as a lens and the telescope are perfectly aligned, the 'gravitational lens' makes a perfectly circular image (an 'Einstein

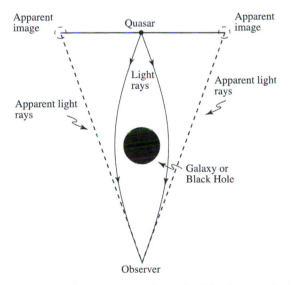

Figure 21. Gravitational lens. The light coming from a far object (a quasar in the figure) is bent by strong gravitational fields existing near a galaxy or a black hole. The Earth observer, by prolonging the rays which come to him, recreates two images of the object as shown. If the object, the observer and the galaxy are perfectly aligned, the system has an axial symmetry and the images fit on a circle that has the object at its centre (Einstein ring).

ring') visible in some photographs taken using various methods in different situations.

The third verification concerned the prediction of general relativity theory that light waves emitted by a source in a gravitational field suffer a change in their wavelength towards longer wavelengths, that is towards the red end of the spectrum, an effect known as a red-shift. This effect has been also verified by studying the light emitted by white dwarf stars which contain matter in a highly condensed state. The gravitational potential at the surface of a white dwarf is much larger than at the surface of our Sun, and the effect predicted by Einstein may be verified. Other consequences of Einstein's gravitation theory have been derived over the following years. One of the most spectacular is the discovery of black holes, collapsed stars whose diameter has shrunken to an infinitesimal fraction of their former state. The light emitted from a black hole never leaves it because it is drawn back by the enormous gravitational field caused by the shrinking. A black hole is, effectively, a point in space with the mass of a normal star. It is now generally accepted that black holes exist, and some observational evidence has been obtained.

During the years 1921–23, Einstein travelled around the United States, Europe and Asia. He was convinced by Weitzmann to engage himself in the Zionist movement. In 1921 he was awarded with the Nobel prize for physics not for relativity theory but 'for his services to Theoretical Physics, and especially for his discovery of the law of the photoelectric effect'. We will describe the photoelectric effect in the next chapter.

An important experimental proof of special relativity was the negative result of an experiment made in 1887 by Michelson and Morley to detect the motion of the Earth with respect to the ether. At the end of the 19th century, it was believed that electric and magnetic vibrating fields of a light wave should necessarily represent the vibration of some medium, and that this medium was the famous ether already introduced by Descartes and Huygens. The electric vibrations of ether had been identified with light and the problem now arose to determine the properties of ether. If for example a light source moves with respect to the ether or an observer moves with respect to it, this motion should be detected. But the Michelson and Morley experiment showed no effect could be detected. The experiment is one of the experimental pillars of the theory of relativity, even if it does not seem to have influenced Einstein in his work of 1905. Probably in fact at that time he was unaware of its existence. In 1921, Dayton C Miller (1886–1941) who was a younger colleague of Michelson at the Case Institute of Cleveland in the United States, performed a similar experiment on the peak of Mount Wilson in California, where an astronomical observatory exists, and declared he was able to observe a small effect of the Earth's motion on the velocity of light. He theorized that at sea level ether is transported by Earth, while at high altitudes a small effect due to the 'ether wind' could be put into light. These results were published a few years later. At this time Einstein was visiting Princeton and, when the news was given to him, he spoke the famous sentence: 'Subtle is the Lord, malicious He is not'.

Miller's result was received with great joy by the enemies of relativity theory. The German reactionary groups had never accepted the reality of the defeat of the military machine of the Kaiser in the First World War and explained it as the result of a 'sinister alliance' between Jews and socialists. Einstein was in their sight: his pacifist and socialist ideas were well known; he refused to sign a declaration of German professors in support of the Belgium invasion and always invoked peace negotiations, even when it seemed Germany was winning the war. Of course the results of Miller's experiment were immediately shown to be wrong by Georg Joss, who did an excellent series of experiments on Jungfraujoch, confirming Einstein's predictions.

Einstein and photon statistics

In 1924 Einstein regained his interest once more in photons in relation to the statistical laws they obey. Satyendranath Bose (1894–1974) at the time physics lecturer at the Dacca University in East Bengal, in 1923 submitted for publication a paper to the prestigious English journal *Philosophical Magazine* in which a new proof of Planck's formula was given. Six months later the editor of that magazine informed him that his paper was rejected, and Bose on 4 June 1924 sent the manuscript to Einstein in Berlin with a letter which began:

'Respected Sir,
　'I have ventured to send you the accompanying article for your perusal and opinion. I am anxious to know what you think of it. . . . I do not know sufficient German to translate the paper. If you think the paper worth publication I shall be grateful if you arrange for its publication in *Zeitschrift für Physik*. Though a complete stranger to you, I do not hesitate in making such a request. Because we are all your pupils though profiting only from your teachings through your writings.'

In fact Bose, already in 1919, together with his compatriot M N Saha (1893–1956) published an anthology of works by Einstein on relativity, one of the first of such collections published in English. Einstein translated the paper and sent it in July 1924 to the *Zeitschrift* in Bose's name, where it was published. He also added a note stating:

'In my opinion Bose's derivation of the Planck formula signifies an important advance. The method used also yields the quantum theory of the ideal gas, as I will work out in detail elsewhere.'

He then sent a postcard to Bose signifying that he considered the work a most important contribution. Bose's paper was the fourth and the last of the revolutionary steps in the history of the old theory of quanta (the other three are Planck's law of 1900, Einstein's light quanta hypothesis of 1905, and Bohr's theory of 1913). Bose's reasoning sets free Planck's law of all the superfluous elements of electromagnetic theory, and obtains the derivation from the bare essentials. He studies the law of thermal equilibrium of particles that have the characteristics of photons, and discovers that these particles obey a new statistical law.

In 1924 Bose received support for a study period of two years in Europe and arrived in Paris in September. He recounts that on the strength of the postcard Einstein had sent him, the German Consulate in Calcutta issued his visa without requiring payment of the customary fee! On arrival in Paris he met Paul Langevin (1872–1946), who suggested to him the possibility of working in Mme Curie's laboratory. Bose recalled the meeting with her with amusement in an interview. Mme Curie (Marie Sklodowska Curie 1867–1934) spoke in English all the time and did not let him say a single word. She told him about another Indian student who worked with her, and had encountered serious difficulties because he did not speak good French. Then she suggested that Bose should concentrate on the language for six months and then come back to her. Bose did not even get the chance to tell her that he had studied French for 15 years!

After this discouraging contact Bose met the de Broglie brothers (Maurice and Louis) and was with Maurice (1875–1960) for some time. However, he was still very anxious to go to Einstein. On 26 October 1924 he wrote Einstein a letter which began:

> 'Dear Master,
>   'My heartfelt gratitude for taking the trouble of translating the paper yourself and publishing it. I just saw it in print before I left India. I have sent you about the middle of June a second paper.... I have been granted study leave by my university.... I don't know whether it will be possible for me to work under you in Germany....'

Einstein translated this second paper and sent it to the *Zeitschrift* which published it. This time, however, Einstein added a remark stating that he could not agree with his author's conclusions, and went on to give his reasons.

The opportunity of a collaboration between Bose and Einstein had already evaporated by January 1925. In July 1924, about the time Bose was finally negotiating the question of study leave with the Dacca University authorities, Einstein was reading a paper before the Prussian Academy where he applied Bose's statistical method to the particles of an ideal gas. The similarity in the statistical behaviour between photons and gas particles found in the paper was further investigated by Einstein in September, deriving important results on their behaviour at low temperature, and in January 1925 Einstein published a second paper where he treated the argument completely, obtaining the statistical laws followed by particles and photons (later called

Bose–Einstein statistics) and turned his attention to other matters. Einstein generalized Bose's theory to a gas of identical particles, atoms or molecules, and predicted that at sufficiently low temperatures the particles would join together in the lower quantum state of the system. This phenomenon is today called Bose–Einstein condensation, and has many unusual properties; for years people tried to obtain it experimentally without success until in 1995, with the help of a laser, a Bose–Einstein condensed state was observed for the first time by two groups at JILA (Joint Institute for Laboratory Astrophysics) in Boulder, Colorado, and MIT. Eric A Cornell (1961–), Carl E Wieman (1951–) and Wolfang Ketterle (1957–) were awarded the physics Nobel prize in 1997 'for the achievement of Bose–Eistein condensation in dilute gases of alkali atoms, and for early fundamental studies of the properties of the condensate'.

After arriving in Berlin, Bose on 8 October 1925 wrote to Einstein asking for a meeting, but Einstein had already ceased to be interested in the argument and was in Leyden. He came back only after several weeks. When finally the two men met, the encounter was very disappointing. As a result Bose obtained a letter allowing him to enjoy some privileges common to students in Berlin, including permission to take books from the university library!

Probably because he had not formally submitted a thesis for a PhD and had not visited England, then the Mecca for Indian scholars, Bose after his return to Dacca was not made a professor. A postcard from Einstein with a single sentence to the vice-chancellor of Dacca University saying that many of those in Europe had benefited from the presence of Bose, later cleared the way to a professorship, and in 1954 Bose came back to Calcutta, where he was born, as a physics professor.

His unsuccessful travel in Europe extinguished his creative vein and he contributed no more to the forefront of physics.

Einstein at Princeton

When Hitler rose to power in 1933, Einstein was travelling in the United States. He did not return to Germany. After a short stay in Belgium, during which he resigned from the Prussian Academy and the Bavarian Academy of Sciences as a protest against the passive behaviour adopted by these Academies when the academic freedom in Germany was suppressed and many scholars and intellectuals were removed from their positions for ideological reasons, Einstein accepted a position at the new Institute for Advanced Study at Princeton in the United States.

The Institute was born from an outstanding donation of Mr Louis Bamberg and his sister Caroline Bamberg Fuld who first had asked Abraham Flexner (1866–1959), the foremost American expert of University instruction, to organize a medical institute, and then, dissuaded by him, had supported his idea to create a new kind of institute where teaching, examination or assigning academic degrees was not mandatory, but where the greatest minds of the time could dedicate themselves to pure research in the most peaceful and free atmosphere, well paid and protected.

The Institute was opened on 20 May 1930 although the inauguration took place three years later. Together with Einstein, there were three full professors: James Alexander (1888–1971), a mathematical expert in topology, John von Neumann (1903–1957), a genial theoretical and experimental physicist who built in Princeton the first computer, and Oswald Veblen (1880–1960), a mathematical expert of differential geometry and topology.

The enrolment of Einstein was made following a series of encounters with Flexner. In the winter of 1932, Abraham Flexner was in California searching for members of the academic body of the new Institute. It was suggested to him to pay a visit to Einstein who was at the California Institute of Technology (CalTech). Einstein immediately liked the idea of the Institute. The situation was rapidly changing in Germany, where since 1920 an anti-Einstein association, a study group formed by the so-called German natural philosophers, was offering money to anybody who would speak against Jewish physics, especially against relativity.

Therefore the two men made a rendezvous in Oxford during the second semester of 1932 and then at Caput, near to Berlin, where Einstein had a small summer house that may still be visited today. Eventually, on 4 June 1932, Albert Einstein accepted the invitation to be the first member of the academic staff of the institute.

During meetings with Abraham Flexner, Einstein asked for an annual revenue of three thousand dollars. 'Could I live on less?' he asked. The agreement already ratified in October 1932 assigned him a stipend of fifteen thousand dollars.

Now there was the problem of an assistant. Einstein wanted Walther Mayer (1887–1948), an Austrian mathematician with whom he had written some papers and wanted him to be enrolled as a professor. Flexner thought that Mayer did not fulfil the prerequisites to be nominated professor, but Einstein was unmovable. So on 17 October 1933, Albert, his wife Elsa, his secretary Helen Dukas (1896–1982) and Walther Mayer, embarked for New York, en route to Princeton.

Here he continued his research, concentrating principally on the creation of what he called a 'unified theory' of fields that should give, so he hoped, a more profound explanation of both gravity and electromagnetism and describe particles as stable regions of a high field concentration.

Einstein did not succeed in his efforts. Besides this research he occasionally came back to the theory of gravity of 1915 and enriched it with new results. In 1932 he collaborated with the Dutch astronomer Willem de Sitter (1872–1934) in the construction of a model of an expanding universe that is still a possible candidate for representation of the large scale structure of the material world.

Einstein, who with his hypotheses of light quanta and of specific heats, contributed in a determinant way to the development of quantum mechanics, never accepted its probabilistic interpretation of nature. At the end of 1927, at the fifth Solvay Congress, a battle with Bohr, Born and Heisenberg fired up. They maintained that indeterminacy is inevitable, but Einstein was not willing to accept this position and presented a series of examples to maintain his point of view. However, Bohr and his supporters replied to all the objections. In 1930 at the sixth Solvay Conference, the last at which Einstein participated, the polemic exploded and eventually in 1935 with the collaboration of two colleagues of the Institute, Boris Podolsky (1896–1966) and Nathan Rosen (1909–1995), Einstein produced a paper of four pages in which the falsity of quantum theory was challenged. The principal arguments, today known as the Einstein–Podolsky–Rosen paradox, represented a small bomb! Bohr, deeply stirred, started immediately to dictate a reply. He understood, however, that the matter was not simple. He started to follow a logical line, then changed his approach and started again. He was not able to determine exactly what the problem was. 'Do you understand what we want to say?' he asked Leon Rosenfeld (1904–1974), the American theoretical physicist who was his assistant on that occasion. Richard Feynman (1918–1988) the American Nobel prize winner for physics of 1965 together with Julian Schwinger (1918–1994) for a calculation method known as Feynman diagrams, with reference to the EPR paradox, said in 1982 'I am not able to define the true problem, then I suspect that there exists no true problem'. Today discussion of this paradox has elucidated certain properties of the quantum mechanical interpretation of nature that were not appreciated at the time, and it has been through using laser light that the problem has been studied giving confirmation of the apparently odd results of quantum mechanics.

In 1936, Einstein had to substitute his favourite ex-assistant Walther Mayer. It seems that Mayer, as soon as he arrived at the Institute, did not hesitate to distance himself from the master. Their collaboration was merely a single paper published in 1934, after which Mayer returned to his interest in pure mathematics. So Einstein in 1936–37 took two new assistants: Peter Bergmann (1915–2002) and Leopold Infeld (1893–1968). He wanted them to continue to work with him also during the following year, but encountered administration difficulties. In the end, Bergmann's post was confirmed but Infeld could not be kept on. Infeld then proposed to write a popular book on the evolution of physics and to publish it with Einstein as co-author. Einstein accepted and Infeld wrote the book during the summer of 1937. When the book, *The Evolution of Physics*, came out in 1938, it generated for its authors much more than the six hundred dollars that Einstein had asked the Institute to pay Infeld.

A joint paper with Nathan Rosen in 1937 presented the solution of his field equations that describe gravitational waves and a famous work published in 1938 in collaboration with Banesh Hoffmann (1906–1986) and Leopold Infeld was dedicated to deriving the equations of motion for particles from the equations of the gravitational field. Even after his retirement in 1945, Einstein continued to work, until his death, on 18 April 1955, at the age of 76 years.

An important characteristic of Einstein's treatment of the fundamental problems of physics was that he questioned also the validity of those concepts and relations that were generally considered as true: in this respect he was a philosopher. According to him, concepts are free inventions and axioms and the fundamental laws of a theory are conjectures; they cannot be deduced or derived inductively from experiments or observation. On the other hand, a theory should allow the derivation of propositions that may be verified experimentally, and in this lies its value. So science requires three human activities: human invention, logical-mathematical deduction and observation or experiment. As Einstein observed, the creation process is guided not only by experience and by pre-existing theories but also by a sense of structural simplicity and mathematical beauty.

CHAPTER 6

EINSTEIN AND LIGHT, THE PHOTOELECTRIC EFFECT AND STIMULATED EMISSION

In June 1905, when Einstein published, in volume 17 of *Annalen der Physik*, his revolutionary paper *Uber einen die Erzeugung und Verwandlung des Lichtes betreffenden heuristischen Gesichtspunkt* (On a heuristic viewpoint concerning the production and transformation of light), everybody had been convinced that light was composed of electromagnetic waves. If one thing was certain, that was it. Einstein, however, planted a doubt, and revealed the dual nature of light: both particle-like and wave-like. Although he was rather critical with respect to Planck's theory, he showed which fundamental conclusions could be derived from it and thus established the crisis of classical physics. At the time Einstein was 26 years old, and his paper appeared in the same volume of the journal in which he published two other fundamental papers: one on statistics, concerning Brownian motion that allowed the old controversy concerning the 'physical' existence of molecules to be cleared up, and the other in which he presented the theory of special relativity. All three papers contribute to making this volume of *Annalen der Physik* one of the most notable in the whole scientific literature.

The photoelectric effect

The paper is today commonly referred to as the Einstein paper on the photoelectric effect; however, it bears a much greater meaning. In it, Einstein deduced, from general considerations of statistical thermodynamics, that the entropy of the radiation described by Wien's distribution law has the same form as the entropy of a gas of elementary particles. Einstein used this argument to reason, from a heuristic point of view, that light consists of quanta, each one having an energy content given by the product of Planck's constant and the frequency of the light, and applied this conclusion to explain certain phenomena, among which was the photoelectric effect. He wrote:

> 'The wave theory, operating with continuous spatial functions, has proved to be correct in representing purely optical phenomena and will probably not be replaced by any other theory. One must, however,

keep in mind that the optical observations are concerned with temporal mean values and not with instantaneous values, and it is possible, in spite of the complete experimental verification of the theory of diffraction, reflection, refraction, dispersion, and so on, that the theory of light that operates with continuous spatial functions may lead to contradictions with observations if we apply it to phenomena of the generation and transformation of light. It appears to me, in fact, that the observations on 'blackbody radiation', photoluminescence, the generation of cathode rays with ultraviolet radiation, and other groups of phenomena related to the generation and transformation of light can be understood better on the assumption that the energy in light is distributed discontinuously in space. According to the presently proposed assumption the energy in a beam of light emanating from a point source is not distributed continuously over larger and larger volumes of space but consists of a finite number of energy quanta, localized at points of space, which move without subdividing and which are absorbed and emitted only as units.'

Einstein uses the words 'energy quantum'. The name 'photon' was introduced much later, in 1926, by the American chemist G N Lewis (1875–1946), one of the fathers of the modern theory of chemical valence.

The production of cathode rays (that is negatively charged particles, identified as electrons) from ultraviolet light, was the way the photoelectric effect was defined at the time. Ironically the phenomenon was discovered in 1887 by Heinrich Hertz, while he was brilliantly confirming the electromagnetic (wave) theory of light with his discovery of electromagnetic waves, and was studied the following year by Wilhelm Hallwachs (1859–1922) who, in particular, showed that certain metallic surfaces, initially deprived of electric charge (discharged), acquired a positive charge when irradiated with ultraviolet light. Later Joseph John Thomson and Philip Lenard (1862–1947) in 1899 showed independently that the effect was produced by the emission of negatively charged particles, electrons, from the metal surface. Because initially the metal had no surplus electric charge, if negative charges were emitted, the positive charges which previously neutralized the emitted negative charges had to remain on the metal. Lenard continued to investigate the phenomenon and presented detailed results in a long paper published in *Annalen der Physik* in 1902. In this paper he reported two important facts: the first one was that in order to obtain electrons from a given metal surface only light of certain frequencies was effective; the second fact regarded the velocity (kinetic energy) of the emitted electrons that did not depend on the intensity of the incident radiation.

Einstein in his work provided an explanation of the photoelectric effect as an example of the application of his theory of light quanta. According to him, the energy of the light wave does not propagate as a wave, but rather as a particle (Einstein called it a 'quantum of energy') that has an energy inversely proportional to the wavelength of the light. The number of quanta is proportional to the light intensity. The more intense a wave is, the more quanta it contains. When a quantum of light collides with an electron in the metal, it gives the electron all its energy and disappears. The electron spends part of this energy escaping from the metal, and keeps what remains as kinetic energy. The intensity of the light beam, being proportional to the number of quanta, has no effect on the energy of the electrons, but determines their total number.

In a letter to his friend Conrad Habicht (1876–1958), Einstein wrote of his paper:

> 'It deals with radiation and the energy characteristics of light and is very revolutionary, as you will see.'

Notwithstanding this declaration, in discussing the physical interpretation of Wien's law and enouncing the concept of a quantum of light, Einstein did not consider that he had broken with tradition. By introducing the quantum of light he applied in a coherent way the statistical methods associated with the theory of radiating heat. However, he called his introduction of the hypothesis of light quanta a 'revolutionary' step because he thought it contradicted Maxwell's electrodynamics that required that radiation to be a continuously radiating energy flow in space.

To understand how Einstein was able to construct such a theory, just in the period in which Planck was trying to demonstrate that his theory of the quantization of oscillators was barely little more than an artifice of calculus, one may consider the personalities of these two scientists who found themselves at that time in very different situations. Planck was a famous mature scientist who wanted to maintain his prestige in the academic community and avoid moving from the scientific theories which were well consolidated at the time. All his efforts were concentrated towards making his discovery to be part of an explanation coherent with Maxwell and Boltzmann theories.

Young and without prejudices or academic obligations, Einstein was then employed in the Swiss Patent Office and therefore had nobody to answer to for his theories. He could take risks. As has been written (Martin J Klein),[1] he 'set himself against the strong tide of nineteenth-century physics and dared to challenge the highly successful wave theory of light, which was one of its most characteristic features'. He argued instead that light can, and for many purposes must, be considered as composed of a collection of independent particles (quanta) of energy that behave like the particles of a gas. This hypothesis of light quanta meant a revival and modernization of the corpuscular theory of light, which had been buried under the weight of all the evidence accumulated in favour of the wave theory during almost a century.

---

[1] Martin J Klein 1963 *The Natural Philosophe* vol. 2, pp. 59–86.

Contrary to what one may suppose, Einstein's hypothesis was not a 'development' of Planck's theory of the blackbody. Einstein was acquainted with Planck's papers but did not share entirely the argumentation. In 1905 he made no use of Planck's theory; he did not use his formula and did not quote his hypotheses. He followed a different path and did not even use the letter h in the expression for the quantum energy of light—that is the product of Planck's constant and the frequency—but employed a combination of constants in which there appeared the constant of the law of perfect gases, Avogadro's number, and a constant already present in the radiation distribution law of the blackbody, given by Wien's formula.

All this however, did not mean the rejection of Planck's ideas nor that light quanta had been invented without previous discussion of the 'elements of energy', but simply that light quanta had not corresponded to a direct derivation or generalization of the elements of energy. Nor was the hypothesis of light quanta motivated by the necessity to explain the photoelectric effect which did not exist as a problem in 1905. Einstein instead was looking for an answer to the general problem, also raised by Rayleigh, as we have seen, to find the reason behind the apparent impossibility to cast blackbody radiation into Maxwell's theory, and it was to confirm the thesis to which he had arrived that he used certain experimental facts, including the results of experiments concerning the photoelectric effect.

The explanation of the photoelectric effect in terms of photons took many years to be completely accepted. The best confirmation of Einstein's theory came from measurements which the American physicist Robert Andrew Millikan (1868–1953) performed during the decade between 1916 and 1926.

> Millikan was born in Morrison, Illinois, and earned his doctorate in physics at Columbia University. Then in 1895 he travelled for one year in Europe to Universities in Berlin, Gottingen and Paris, meeting with Max Planck, Walther Nernst, and Henri Poincarè. In 1896 he was assistant to Albert A Michelson at Chicago University where he was appointed professor in 1910. In 1921 he moved to the California Institute of Technology. In 1923 he was awarded the physics Nobel prize 'for his precision measurements for the electronic charge and Planck's constant'.

Millikan, who initially did not trust Einstein's theory, provided the best proofs of its validity and was awarded the physics Nobel prize partly also for these results. The final evidence came later when the American physicist Arthur Holly Compton (1892–1962) found in 1922 that x-rays were scattered by free electrons as if they were particles of energy hf (f being the frequency of the radiation) and momentum hf/c, as Einstein had predicted. In particular the scattered quantum had a different frequency from that of incidence and this frequency changed with the scattering angle (the Compton effect; Compton was awarded the Nobel prize for this in 1927), facts that were not possible to explain with the wave theory. At that time, however, Einstein's hypothesis of light quanta was already fully accepted.

At the beginning, however, the contemporary scientific world did not believe Einstein's theory of the photoelectric effect. In 1913, in a letter which proposed Einstein for membership of the Prussian Academy and for a research professorship and which extolled his work and his genius, Max Planck wrote: 'That he may sometimes have missed the target in his speculations, as for example, in his hypothesis of light quanta, cannot really be held against him'.

A few years later, in 1916, describing his experimental confirmation of Einstein's equation for the photoelectric effect, Millikan wrote of the same hypothesis: 'I shall not attempt to present the basis for such an assumption, for, as a matter of fact, it had almost none at the time'.

In the end Einstein was awarded the Nobel prize in 1921 not for his relativity theory but just for his theory of the photoelectric effect.

In 1906, Einstein took a deeper look at the way Planck had derived the blackbody law in a study entitled *Theorie der Lichterzeugung und Lichtabsorption* (On the Theory of emission and absorption of light) and concluded:

> 'We must therefore regard the following law as the basis of Planck's quantum theory of radiation. The energy of an elemental resonator [oscillator] can only assume values which are integral multiples of [the energy of the quantum of light]; the energy of a resonator changes in jumps by absorption or emission in integral multiples of [the same quantity].'

With those words Einstein focused sharply on what he considered to be the main hypothesis of Planck's radiation theory, namely, the fact that the resonators in a cavity change their energy only by finite amounts, that is, discontinuously in discrete steps. Two years later, Lorentz reached the same conclusion, that Planck had introduced an essentially new hypothesis which contradicted the usual laws of electrodynamics.

In 1909, four years after his work on the photoelectric effect, Einstein published a paper in which he demonstrated that Planck's radiation law implies that radiation exhibits a combined wave and corpuscular nature. This result was the first clear indication of the so-called wave-particle duality that later will be largely considered in quantum mechanics.

In retrospect it is interesting to observe that in the quarrel of the 17th century regarding the wave or particle nature of light, made between giants (Newton and Huygens), both opponents had gathered in their own way one side of a two-sided problem.

Stimulated emission

Leaving the development of quantum theory—which received its full recognition during the first Solvay Congress, organized by Walther Nernst in 1911 with the purpose of provoking an open discussion on the 'crisis' introduced in physics by quantum ideas and with the financial support of the Belgian scientist Ernest Solvay (1838–1922) who invented an industrial method for the preparation of soda—we now return to Einstein's studies of light.

Einstein was greatly drawn to the problem of the nature of light, and in 1915–16 published a paper '*Strahlungs-Emission und -Absorption nach der Quantentheorie*' that is fundamental and pivotal to our history. He had continued to brood over Planck's blackbody theory and the in-some-sense artificial way with which Planck had solved the problem by introducing the concept of quantization of energy. Then in 1916 he published a new, extremely simple and elegant proof of Planck's law of radiation and, at the same time, obtained important new results concerning the emission and absorption of light from atoms or molecules. In this paper, for the first time, the concept of stimulated emission, which is fundamental to the laser effect, is introduced. He skilfully combined 'classical laws' with the new concepts of quantum mechanics, which were at the time growing under the guidance of Bohr.

Einstein considered molecules in a vessel. According to the postulates of Bohr then already developed, each molecule can have only a discrete set of states, each one with a defined energy. If a large number of such molecules belong to a gas at some temperature, the probability that one molecule finds itself in one particular state can be derived by applying the laws of statistical mechanics developed by Gibbs, Maxwell and Boltzmann. Einstein assumed that the molecules exchange energy with the radiation present within the volume through three processes. The first one is the process that today we call 'spontaneous emission', which occurs if the molecule does not stay in the lower energy state but in some higher energy state; then it shall decay to the lower energy state emitting a photon that has exactly the energy corresponding to the energy difference between the two states (figure 22a). This de-excitation process, that is the process described by Bohr for a molecule or an excited atom to jump into a state with lower energy, is a process that Einstein assumed would happen in a casual way, similar to the casual way a radioactive atom disintegrates in time. The second process may be considered as the reverse of the previous one, and is the absorption process. A molecule that finds itself in some energy state may be raised to a higher energy state if it is struck by a photon that has just the energy corresponding to the energy difference between the two states (figure 22b). This process had also been considered by Bohr. In this case, the photon disappears (is absorbed) and the molecule takes all its energy to raise to the higher energy level. The third process was introduced by Einstein here for the first time and is today called 'stimulated emission'. According to this process, if a molecule finds itself in a higher energy level and is struck by a photon, it may decay to a lower energy state, if the photon has exactly the energy corresponding to the energy difference between the two states. When

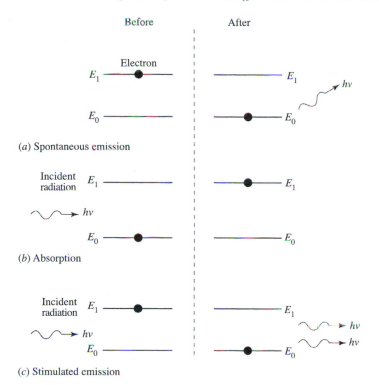

Figure 22. Upper side (a). An electron sitting on an upper level decays spontaneously to the lower level (spontaneous emission) by emitting the energy difference as a photon which is emitted randomly. In the middle (b) an electron is pushed to jump from a lower to an upper level by a photon which has an energy equal to the energy difference between the two levels and which is absorbed (absorption process). The lower diagram (c) illustrates the stimulated emission in which a photon with the right energy (that is an energy equal to the energy difference between the two levels) strikes the electron which sits in the upper level and stimulates it to jump to the lower level by emitting another photon identical to the one which has stimulated the process.

decaying to the lower energy state, the molecule emits a photon that has exactly the same energy of the first one, while the first photon continues its journey unaffected, happy to have 'stimulated' the molecule to de-excite (figure 22c).

If we now assume that the molecules may interact with radiation through these three processes, and that the interaction must not change the energy distribution that depends only on temperature and is given by the Maxwell–Boltzmann law, Planck's law is immediately derived, together with certain relations that link coefficients that describe the three processes. These coefficients are today called Einstein coefficients and enable the probabilities of transitions between states to be derived. Bohr's quantum theory did not give any indication of the laws governing such transitions, the concept of transition probability originating in Einstein's paper.

Einstein's derivation of Planck's distribution law through the introduction of probability coefficients for absorption, spontaneous and stimulated emission allowed the processes to be linked via these coefficients. Einstein was not able to express them in terms of the characteristics parameters of the atom. Such an expression would be given more than ten years later by P A M Dirac, utilizing at the time fully developed quantum mechanics. However, the expressions found by Einstein established that the absorption and stimulated emission coefficients were equal and that the ratio between the spontaneous emission and absorption is inversely proportional to the cube of the wavelength. Because the spontaneous emission probability can be experimentally measured, Einstein's formulae could be verified by comparison of the intensities of absorption and spontaneous emission spectral lines.

Another important result established in Einstein's work refers to the fact that when the atom or molecule exchanges energy with the radiation, by absorbing or emitting a quantum of light, there is also an exchange of momentum, exactly as happens when two billiard balls collide. The atom which emits a photon in some direction, recoils in the opposite direction, in exactly the same way as a shotgun recoils.

Some time later, in 1923, the German physicist Walther Wilhelm Georg Franz Bothe (1891–1957) used Einstein's theory on emission and absorption to show, among other things, that the light quantum emitted in the process of stimulated emission, besides having exactly the same energy as the stimulating quantum, travels in the same direction, that is it has the same momentum of the incident quantum. This type of behaviour is exactly what is needed in order to have an amplification process, because it means—by using a classical language—that a wave travelling in a medium where excited atoms or molecules are present, is reinforced by the wave emitted in each stimulated process it produces, i.e. it is amplified.

For about 30 years, however, the concept of stimulated emission was used only theoretically, and received only marginal attention from the experimental point of view. Even in 1954 W H Heitler (1904–1981) writing a monograph that is a classic on the quantum theory of radiation gave very limited space to this argument.

The role of stimulated emission in the theory of light dispersion
By using Einstein's results, theoretical physicists were able to build up quantum theories of light diffusion and dispersion.

As we said, speaking of the refraction of light by a prism, light rays that correspond to different colours incline with different angles because they travel at different velocities. To describe this phenomenon it is convenient to introduce a quantity that takes the name of refractive index and represents the ratio between the light velocity in vacuum and the light velocity in the medium. By using this quantity the refraction law may be stated by saying

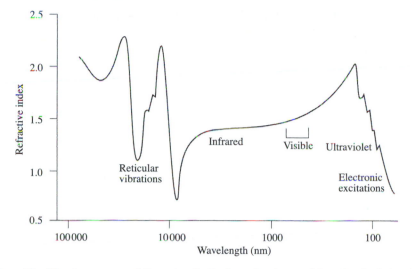

Figure 23. The phenomenon of dispersion. In the figure the change of the refractive index of a colourless glass is shown as a function of wavelength expressed in nanometres (1 nm = 10^{-9} m).

that the sine of the angle of incidence divided by the sine of the angle of refraction is equal to the ratio of the refractive index of the second medium to the refractive index of the incidence medium.

   The phenomenon by which the propagation velocity of light (that is the refractive index) depends on the wavelength takes the name of light dispersion (figure 23). The reason why light of different colours travels in the same medium with different velocities was discovered by a study of the way electrons in atoms emit light. The simplest model that can be used is the one in which the electron in an atom moves forwards and backwards in a regular motion, as does for example the pendulum of a clock. A motion of this kind in physics is denoted as periodic motion. During its motion the electron experiences accelerations and, therefore, according to Maxwell's equations it should emit radiation. All this may be represented by a simple model in which the electron is elastically bound to the atom as if it were connected by a spring (harmonic oscillator). This model had already been used to describe the emission of radiation in the blackbody. It was now used to explain the emission and absorption of electromagnetic radiation by matter.

   To explain why an atom is able to emit many frequencies, one may assume it is composed of many oscillators able to emit or absorb certain frequencies that are the frequencies that the atom experimentally is found to absorb. In this way P Drude, W Voigt (1850–1919) and later H A Lorentz elaborated a theory of dispersion that was in a good agreement with the experiments and gave a satisfactory explanation of dispersion and absorption of light. Studying mathematically the response of the oscillators to the electric field of

109

the wave, the refractive index and its dependence on wavelength can be derived. The result is interesting and shows that at wavelengths far from those at which the atom absorbs, the refractive index is unity, that is the light propagates with the same velocity as in vacuum, being unaffected by the medium. However, as the wavelength approaches the wavelength at which the atom can absorb, the refractive index decreases (as the absorption increases) and after reaching a minimum starts to increase and is again unity at the wavelength at which the atom absorbs (but we cannot perceive this because all the light is absorbed). Then, as the wavelength continues to increase, the refractive index increases further, reaches a maximum and then returns to unity very far from absorption. This is exactly what is found experimentally. The behaviour of the refractive index between the minimum and maximum, very difficult to measure because it corresponds to a region of large absorption, is indicated as anomalous dispersion, because in this region the refractive index increases with increasing wavelength instead of decreasing.

The classical equations obtained by calculation were in fairly good agreement with experiment and gave a satisfactory interpretation of dispersion and also absorption. However, when Bohr's theory of stationary states superseded the classical theory of elastically bound electrons, these formulae, notwithstanding their *de facto* validity, completely lost their theoretical justification. The first attempts to formulate a dispersion theory in terms of quantum mechanical concepts, made by P Debye (1884–1966), A Sommerfeld (1868–1951) and C J Davisson (1881–1958), were unsatisfactory, mainly because now in the new model of the atom, by applying the electric field of the light wave, the oscillation was the one produced when the electron was perturbed from its stationary orbit. In this case it started to oscillate around its equilibrium position with a frequency that obviously was very different from the one corresponding to the transition from one orbit to the other.

The first correct step towards the formulation of the quantum mechanical interpretation of dispersion was taken by Ladenburg. Rudolf Walther Ladenburg plays an important role in our history: as we shall see, he came very close to discovering amplification by stimulated emission, which is at the basis of the working action of lasers.

Ladenburg was born in Kiel, Germany, on 6 June 1882 and died in Princeton, New Jersey, on 3 April 1952. He was the third of three sons of the eminent chemist Albert Ladenburg. After school in Breslau, where his father, the author of a number of important research papers in organic chemistry, was chemistry professor at the University, Ladenburg in 1902 went to Munich, where he took his degree in 1906 with a thesis on viscosity, under Roentgen. From 1906 to 1924 he was at Breslau University first as a Privatdozent and from 1909 as Extraordinary Professor. During that period he performed studies on the photoelectric effect and verified that the energy of photoelectrons is independent on the light intensity, but proportional to frequency.

He married in 1911 and three years later served in the army as a cavalry officer, but later on, during the 1914–18 war he carried out research on the use of acoustic signals to identify targets (sonar). In 1924, on the invitation of the director, the Nobel prize winner for chemistry (1918), F Haber (1868–1934), he went to the Kaiser Wilhelm Institute in Berlin, the prestigious Institute where Einstein also served, and where he stayed as Head of the Physics Division until 1931 when he went to Princeton to succeed Karl Compton (1887–1954), the brother of Arthur, to the physics chair.

After the First World War, Ladenburg was looking for some way to connect the Bohr postulates on emission and absorption of radiation from atoms with the model of harmonic oscillators. Although he does not make explicit mention of it, he assumed that when the atom is perturbed, the electron does not oscillate around its orbit as one should expect by applying the classical concepts, but it decays on a lower level in agreement with Bohr's model and that this process could be described classically as if the electron was a small harmonic oscillator that oscillates just at the transition frequency.

The introduction of Einstein's coefficients of absorption, spontaneous and stimulated emission allowed him to propose a theory capable of explaining the optical behaviour of matter. He started in 1921, by deriving an expression that allowed him to find for every atom how many electrons participated in the optical phenomena (this number he called the number of dispersion electrons) in terms of the Einstein coefficient that described the spontaneous emission from an excited atom. He obtained this number by calculating the energy emitted and absorbed by a set of atoms in thermal equilibrium with the radiation and by using the model of the oscillator on the one hand and of Bohr's quantum theory on the other. According to Bohr's correspondence principle, the result of these two calculations, although so different, must be the same. Therefore by equating the results, a relation is found between the number of electrons that participate in the emission or absorption and the Einstein coefficient that describes the spontaneous emission from atoms. The number of electrons that participated in these processes could be derived from experimental measurements of emission, absorption, anomalous dispersion, and so on, and therefore from these measurements the probability with which the different transitions occurred could be derived. Ladenburg applied this result to measurements he performed in hydrogen and sodium between 1921 and 1923.

In 1923, together with Fritz Reiche (1883–1963), he developed a relation that links the refractive index to the wavelength and the Einstein spontaneous emission coefficient. The formula is, however, incomplete because in it does not appear a term due to the effect of stimulated emission. This term was introduced by Kramers and Heisenberg. The fundamental step was made in 1924 by Kramers who modified the formula found by Ladenburg and showed that a term should appear that took explicitly into account the stimulated emission.

Hendrik Anthony Kramers was born on 17 December 1894 in Rotterdam, where his father was a physician. He studied at Leyden University, principally with P Ehrenfest (1880–1933), who in 1912 had succeeded H A Lorentz. In 1916 Kramers went to Copenhagen to work with Niels Bohr. When, in 1920, the Bohr Institute of Theoretical Physics opened, Kramers was at first Assistant, and then in 1924, Lecturer. In 1926 he accepted the theoretical physics chair at Utrecht and in 1934 he returned to Leyden as the successor of Ehrenfest, who committed suicide in September 1933. From 1936 until his death on 24 April 1952, Kramers taught at Leyden and paid a number of visits to other countries, including the United States.

In Copenhagen he worked on the dispersion problem. In 1924 he wrote an expression in which stimulated emission was taken into account. The basic idea of his work was that dispersion must not be calculated by considering the real orbit of the electron interacting classically with the electromagnetic wave. It must instead be derived by substituting for the atom a set of hypothetical oscillators whose frequencies correspond to the jumps between the atomic stationary states of Bohr's model. Each oscillator corresponds therefore to one of the possible atomic transitions. The set of these fictitious (virtual) oscillators was called by Alfred Landé (1888–1975) the 'virtual orchestra'. The virtual orchestra is therefore a classical formal substitution for the radiation and so indirectly becomes the representative of the quantum radiator itself.

Of course in this way it is possible to have positive terms that correspond to a transition from a state of lower energy to one of higher energy with the absorption of a photon, and negative terms that correspond to the reverse transition from a higher to a lower state with the emission of a photon. A negative contribution adds to dispersion that we will indicate as 'negative dispersion', due to the emitting oscillators, and is analogous to the negative absorption represented by Einstein's stimulated emission coefficient, i.e. 'light waves of this frequency, passing through a great number of atoms in the state under consideration, will increase in intensity', as Kramers wrote in a paper in 1925.

Using a highly sophisticated spectroscopic technique, Ladenburg and his collaborators, in the years between 1926 and 1930, studied the effect of negative dispersion. In one of these studies, performed in collaboration with H Kopfermann (1895–1963), Ladenburg investigated the dispersion of gaseous neon near its red emission lines. Neon was excited in a glass tube by means of an electric discharge, more or less as it is done today in luminous signs, and the two scientists measured dispersion as a function of the intensity of the discharge current. They found that by increasing the current above some value, dispersion decreased (that is the difference from unity of the refractive index was falling) and correctly observed that, because the number of atoms in the higher energy state increased, the effect of negative dispersion could explain the decrease of dispersion. These experiments gave the first experimental proof of the existence of negative terms in the dispersion equation. If the measurements had continued systematically, amplification by stimulated emission probably could have been obtained at that time.

Other researchers studied the effects of stimulated emission. J H van Vleck (1899–1980), one of the most prominent American theoretical physicists among the founders of the modern theory of solids and of magnetism in particular, who obtained his PhD at Harvard with the first American thesis on quantum mechanics in 1922 and was awarded the Nobel prize in 1977 with N F Mott and P W Anderson 'for his quantum mechanical description of the magnetic properties of matter'—and the American R C Tolman (1881–1948)—a relativity and statistical mechanics scholar, the discoverer of an effect that demonstrates the existence of free electrons in metals—observed that stimulated emission, called 'induced emission' by van Vleck, may lead to a negative absorption, and Tolman wrote that '... molecules in the upper quantum state may return to the lower quantum state in such a way as to reinforce the primary beam by "negative absorption"'. Tolman deduced 'from analogy with classical mechanics' that the negative absorption process 'would presumably be of such a nature as to reinforce the primary beam'. After having so clearly prepared the basis for the invention of the laser, Tolman said that, for absorption experiments usually performed, the amount of negative absorption can be neglected.

The reason why scientists considered that the phenomena connected to stimulated emission could not yield relevant experimental effects may be found in the conclusions that can be obtained by applying the Maxwell–Boltzmann law which was derived at the end of the 19th century and that allows the probability that a system at equilibrium has a certain energy to be obtained. This law, applied to our case, establishes that at thermal equilibrium, by assuming a set of atoms that may exist in the ground state or in an excited state, the number of atoms in the excited state is always much smaller than the number of atoms in the fundamental state. In nature, all physical systems are at thermal equilibrium or differ very little from it, and come back to it very rapidly. Therefore, in the case of atoms, one has to expect that the number of excited atoms will always be very small in comparison with those in the fundamental state. It is therefore reasonable to expect that the effects due to stimulated emission, that require the presence of excited atoms, will be very small.

Later on, in 1940, the Russian V A Fabrikant in his doctoral dissertation observed that if the number of molecules in the excited state could be made larger than that of molecules in the fundamental state, radiation amplification could occur. However, this thesis was not published and it seems it had no consequence even in Russia. The suggestion was known only when, after the invention of the maser, Fabrikant obtained a Russian patent.

Eventually, in 1947, W E Lamb Jr (1913–) and R C Retherford (1912–) wanted to check the accuracy of the predictions of Paul Dirac about the energy levels and spectral lines of hydrogen. Dirac's prediction was that the hydrogen atom had two possible energy states with equal energy. In a famous experiment made by studying a discharge of hydrogen, the two researchers revealed that there was a minute difference in these energy levels. This 'Lamb shift' showed that a revision of the theory of the interaction of the electron with electromagnetic radiation was needed and resulted in Lamb being awarded the Nobel prize for physics in 1955, which he shared with Polykarp Kusch. In an appendix of a paper on their work published in 1950 Lamb and Retherford, discussing their results, observed that in their experiment the conditions could be created to achieve population inversion (i.e. more excited atoms than the ones in the fundamental state). However, they concluded that their calculations were too optimistic and they did not refine their argument further. Later, Lamb wrote that, at the time, the concept of negative absorption and the earlier studies were new to them, and that in any case their interest was principally directed at the study of the properties that brought him to the Nobel prize, and therefore they did not examine closely the aspects of the problem connected to stimulated emission.

CHAPTER 7

MICROWAVES

We wish now to go back in time to the end of the 18th century, immediately after the publication (1873) of the famous *Treatise on Electricity and Magnetism* by Maxwell.

Notwithstanding the progress made by Maxwell and his first followers in the theory of electromagnetic oscillations, the bridge between classical electrodynamics and the theory of light had not yet been made, except for the brilliant intuition of Maxwell that electromagnetic and light waves were of the same nature. The Irish physicist George Francis Fitzgerald (1851–1901) laid the first stone in 1882 by arguing that, if the unification indicated by Maxwell is valid, it ought to be possible to generate radiant energy by purely electrical means. 'It seems highly probable that the energy of varying currents is in part radiated into space and so lost for us' he stated, seizing only the negative side of the phenomenon, and described methods by which the radiant energy could be produced. He remarked, however, that the difficulty would lie in detecting such waves when they were produced, because suitable detectors did not yet exist.

The experimental discovery of electromagnetic waves

In parallel with the theoretical elaboration of Maxwell equations, experimental research was performed on the generation of electrical oscillations produced by means of the discharge of a common condenser in an electric circuit and detected as an oscillating current in the same circuit. Since 1847, Hermann von Helmholtz had affirmed that in some cases the discharge of a condenser must assume an oscillatory character. The physicist William Thomson in 1853 gave a mathematical formula to establish for which values of the parameters of the components of the circuit the oscillation was obtained.

By working with oscillating circuits of this kind, Heinrich Hertz, a young German then unknown, succeeded in generating and detecting electromagnetic waves.

Heinrich Hertz (1857–1894) was born in Hamburg the son of an attorney, then senator. A very bright student, he did equally well in the humanities and the sciences, and showed great manual skill in designing and constructing scientific apparatus. Hertz was supposed to have followed the tradition of law in the family, but from the age of ten years he was interested in natural sciences and, after attending a number of schools, he decided to study engineering at Dresden Polytechnic in 1876. Being 20 years old, the following year he had to do his military service. After that, he decided to finish his engineering studies in Munich, but soon gave it up for physics. In 1878 he enrolled at the University of Berlin to work under Helmholtz and Kirchhoff and earned his PhD in 1880. In Berlin in 1870 Hermann von Helmholtz had come from Heidelberg, moving from a physiology to a physics chair. For many years Helmholtz had been interested in the physical properties of organisms and biological processes and in particular to the process of perception. These studies had convinced him that a complete description of the processes concerning the nervous system required the understanding of the energy exchanges in living bodies and in this analysis an important role was played by thermodynamics—to which he had already made important contributions, participating in the enunciation of the principle of energy conservation—and electricity. When he arrived in Berlin he started a series of investigations into electricity, and Hertz who arrived in 1878 worked in the frame of this research. He was so fortunate as to attract the attention of Helmholtz who, in the same year in which Hertz earned his doctorate, appointed him his assistant. In 1883 Hertz became Privatdozent at Kiel, with the help of Kirchhoff, and in 1885 was appointed physics professor at Karlsruhe. This university had asked for someone who could teach electrical technologies. The recent developments of energy transmission, the electric light and the applications of electricity, had made electricity the principal technology of the time, and the works Hertz had already made on the subject, together with the support by Helmholtz, won him the position. He died very young of chronic blood poisoning, the same year in which his mentor Helmholtz died.

As often happens, Hertz arrived at the discovery of electromagnetic waves without looking for it intentionally.

In 1879 the Berlin Academy of Sciences offered a prize for research on the problem of establishing experimentally a relationship between variable electric fields and the response of materials to these fields (polarization). At this time, Hertz was involved in electromagnetic research at the Physikalisch-Technische Reichsanstalt of Berlin, and his teacher von Helmholtz drew his attention to the problem. Initially Hertz approached the study of electrical oscillations produced using a Leyden jar (a special kind of electric condenser), but soon concluded that only effects 'lying just within the limits of observation' could be produced, and therefore put the problem aside, only to return —and solve it—nine years later, in 1888, as part of his classic experimental work on electromagnetic waves. A few years later, in 1886–87, during some experiments, Hertz found that if an open circuit, formed of a piece of copper wire bent into the form of a rectangle so that the ends of the wire were separated only by a short air-gap, was put near to a circuit through which the spark discharge of an induction coil was taking place (we will call it the primary circuit), a spark passed through the air-gap of the open circuit. He rightly interpreted the phenomenon by showing that the bent wire (we will call it the secondary circuit) was of such dimensions as to make the free period of electric oscillations in it nearly equal to the period of the oscillations of the primary circuit.

The discovery that sparks may be produced in the air-gap of a secondary circuit, provided it has the proper dimensions for resonance, provided a method of observing electrical effects in air at a distance from the primary disturbance: the detector required by Fitzgerald to observe the propagation of electric waves was now at hand.

Unknown to him, Hertz had been preceded, about seven years earlier, by David Edward Hughes (1830–1900), who had shown that electric sparks could be detected at distances up to about 500 m by a microphone contact (essentially what was later called a 'coherer') to which a telephonic earphone was connected. He rightly claimed that the signals were transmitted by electric waves in air, and in the years 1879–80 demonstrated these experiments before the President of the Royal Society, Sir George Stokes and Mr W H Preece, the Electrician to the Post Office. Unfortunately they came up with another explanation of the phenomena and Hughes, discouraged, did not publish his results until long afterwards.

After a short interruption during which Hertz was busy studying the effect of irradiation with ultraviolet rays upon electric discharges, performing thus the first observations of the photoelectric effect, in 1888, by improving the circuit that produced the sparks and by detecting the effect with his secondary circuit, Hertz was able to demonstrate that electromagnetic waves were emitted.

Also Oliver Lodge (1851–1940), early in 1888, had discovered electromagnetic waves. He demonstrated their propagation and reflection along wires and performed accurate measurements of their wavelength. However, instead of publishing his results immediately, he went on vacation to the Alps, believing his experiments would make an impression at the next congress of the British Association for the Advancement of Science to be held in September. At the Congress, Fitzgerald, to whom the work of Lodge was unknown, announced boisterously that an unknown German, Heinrich Hertz, had generated and detected electromagnetic waves in air.

To generate his sparks Hertz used different experimental configurations. One of them is shown in figure 24. It consists of a condenser of a particular shape made by two metallic spheres of about 30 cm in diameter at the end of a straight copper wire. The centres of the spheres were 1 m apart. The wire was cut at the centre with two knobs at the extremities of the cut. The distance between the two knobs (a few centimetres in diameter) could be adjusted to a value typically of the order of 1 mm. Dimensions were calculated so as to produce waves that could be detected by a secondary circuit of suitable dimensions. In order to excite the circuit a Ruhmkorff

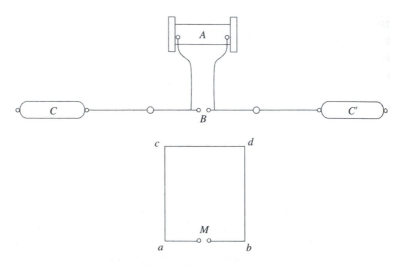

Figure 24. In the upper part the Hertz oscillator and in the lower part the resonator. *A* is a Ruhmkorff coil, *B* are the two spheres between which the spark fires, *C* and *C'* are the two larger conductors that are charged by the coil. The circuit *abcd* is the Hertz resonator and *M* are the two spheres between which the spark is seen. (From H Hertz 1893 *Electric Waves*, reprinted by Dover Publications, New York, 1962, p 40).

coil was used that charged the two spheres with electricity of opposite sign, until a spark fired and travelled the short distance between the two knobs. In this way the system discharges through a series of oscillations whose frequency is determined essentially by the dimension of the spheres and by their reciprocal position. The oscillations cease when the energy associated with the initial charge is irradiated into space in the form of electromagnetic waves. By the end of the year, Hertz had demonstrated the similarity of electromagnetic waves to light, showing they could be reflected, propagated in a straight line after passing through an opening in a screen, by suffering diffraction, and had other properties similar to light.

The waves generated by Hertz in his first experiment had a wavelength of a few metres but soon, in the same year, he was able to generate waves of the order of 10 cm. Paradoxically, Hertz did not realize the potential practical applications of his discovery. When a German technician suggested to him that the waves he had discovered could be used for wireless telegraphy, he discouraged the idea, maintaining that currents in his resonator oscillated millions of times per second and could not be reproduced in a telephonic device that responded only to currents that varied a few thousands times per second.

Hertz's experiments were continued by the Italian Augusto Righi (1850–1921), a physics professor at Bologna University, who was able to generate waves of a few centimetres by detecting them with a resonator that he had made with a rectangular piece of tinfoil, fixed on a glass support.

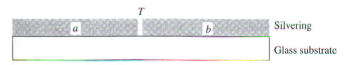

Figure 25. The Righi resonator was made by a thin silvered glass (a normal mirror). The silvering of the glass presents a thin cut *T* that divides it into two metallic conductor parts *a* and *b*, insulated from one another. Exposing the resonator to the waves, the cut *T* is crossed by a spark visible in the dark.

The tin-foil was interrupted with a very thin cut in which the spark was created (figure 25).

> Righi was born in Bologna, and after attending a technical school enrolled in the Mathematics Faculty at Bologna University in 1867, changing for Engineering, and obtaining his degree in 1872. For his dissertation in civil engineering (the only one that existed at the time) he built an electrostatic machine for measuring very small electric charges that may be analogous, however, on a reduced scale, to the famous particle accelerator built by Robert Jemison van de Graff (1900–1967) in the 1930s. A pupil of Antonio Pacinotti (1841–1912) when he was teaching to the Technical Institute, Righi succeeded his master in the teaching of physics at the same school from 1873 to 1880. Here, by utilizing the school's laboratory, he improved the telephone, developed during this period by Alexander Graham Bell (1847–1922), assembled and patented the first microphone with conducting powders and the first loud-speaker, that he presented at the 1878 Universal Exhibition of Paris. He started his university career initially in Palermo (1880–85) and then in Padua where he remained until 1889 when he came back to Bologna as physics professor until his death. Here he carried out studies on the optical properties of electromagnetic waves. In Bologna he established a modern physics Institute. Among the contributions given by Righi, we remember he was the first to observe the hysteresis cycle in magnetic materials. He considered at length electrical discharges in gases, magnetic effects on light and was one of the first to study the photoelectric effect, building a photoelectric pile.

With an oscillator built by himself and named the 'three-sparks oscillator', Righi made a famous series of experiments that were summarized in the book *Ottica delle Oscillazioni Elettriche* (1897) (*Optics of Electric Oscillations*) in which he demonstrated the exactness of the electromagnetic theory.

Marconi and the radio

The Italian Guglielmo Marconi (1874–1937) succeeded in beating a considerable number of great scientists of his time in developing a wireless telegraph system. The 'Marconi affair' may be considered a good example of certain kinds of inventions. In fact the technical devices—oscillators and resonators—utilized by Marconi, even though he improved them systematically, were not original. He was able to take advantage of an exceptionally

119

favourable situation in which nearly all the elements essential for wireless telegraphy were already available, and was able to grasp the potential of a technique which nobody had already considered, pursuing his purpose with great firmness. The fact that all the essential elements for assembling the wireless telegraph were already known, and that many researchers had performed experiments involving electromagnetic waves in one way or other, after the first Marconi experiments stimulated a race to claim ownership in which nearly every country challenged the legal right to possession of the 'invention'. Actually the 'invention' consisted not of the discovery of a new effect or of the invention of some new device, but in the clever use of what already existed, assembling everything in an improved way, and in achieving with unshaken determination a technology considered impossible by many people. At this time, in fact, people believed that electromagnetic waves, irrespective of their wavelength, all behaved like light and propagated in a straight line. Therefore, transmitter and receiver needed to see each other, and transmission distances exceeding a few kilometres were therefore impossible to cover. Marconi did not care for these predictions, partly because being mostly a self-taught man he had not the theoretical basis and the mental habit of the researcher to be influenced by these considerations, and went on his way with a pure pioneer spirit of adventure.

> Guglielmo Marconi was born the second son of a well-to-do land-owner and a rich Irish lady. Although the principal house of the family was at villa Grifone di Pontecchio, 15 km from Bologna, the mother and sons were accustomed to passing long winter periods, principally in Livorno and Florence. During his first studies, the young Marconi began to take a growing interest in experimental sciences, focusing his attention principally on electricity. From the beginning he was interested in practical applications. In 1892 he built a device for the detection of atmospheric electricity that was raised onto the roof of the house at Livorno. He also tried to build an accumulator to supply a small network for the illumination of a rest house the family owned at Cattolica. From the notes written in his notebook, it appears that the young man could rely on financial means more than fair for a young amateur, and had a quite clear understanding of the commercial value of inventions. His mother arranged for her son to visit frequently Augusto Righi, and certainly these visits were very useful to him, even if Righi did not take the ideas of the young amateur scientist very seriously, being of the opinion that if he had really wanted to devote himself to research, he should first of all finish his studies. Certainly he gave him advice and suggestions and allowed him to assist in some experiments and to attend the library, which placed him in a position in which he could follow all the scientific literature on the subject.

In 1894–95, in his house near Bologna, Marconi started a series of experiments to employ electromagnetic waves to establish distant communications, in a wireless telegraphy system, using as a generator a Righi oscillator and as a detector a coherer with metal filings.

The coherer is a device made of a small evacuated glass tube of a few milli-metres diameter, in which two small silver cylinders are separated by a short distance and connected externally via two platinum wires soldered in the glass. Between the two cylinders a small quantity of metal filings is placed (nickel and silver) with medium sized grains. The filings present a large resistance to current flow, but when they are subjected to an electromagnetic wave the grains cohere (hence the name), that is they align in the direction of the current and the resistance greatly decreases. This phenomenon had been already discovered by Temistocle Calzecchi Onesti (1853–1922) in 1884 and was used by Marconi who succeeded in improving the performance of the device with a better choice of materials and by introducing a hammer that, after every electromagnetic pulse, re-established automatically the resistance of the filings and prepared the device for the detection of successive pulses. Also the French physicist Edouard Branly (1844–1940) had observed in 1890 that metal filings placed in a glass tube changed between an insulating state to one of conduction when an electric spark fired at some distance. Probably this property would have interested nobody if the tube containing the filings, later called coherer by the English physicist Oliver Lodge, had not become the essential element of the first radiotelegraphic detectors, thanks to its use by Marconi, so causing a war of claims for ownership among Calzecchi Onesti, Branly, Lodge and Popov, a Russian physicist who used it for the same purpose.

After having more or less repeated the experiments known at the time that allowed signals to be propagated over distances of a few metres, Marconi discovered, in August 1895, that to observe appreciable effects at distance it was necessary to connect the extremities of the oscillator and of the detecting circuit on one side to a conductor underground (earth) and on the other side to an insulated conductor (antenna) as high as possible above the ground. The brilliant invention of the antenna–ground system allowed intelligible telegraphic signals to be received over distances up to 2400 m. Surprisingly the signals were detected on the other side of a hill interposed between the detecting and transmitting systems. In one of these experiments Guglielmo's brother Alfonso confirmed the detection of the signal on the other side of the hill with a rifle shot that has entered the legend of the Marconian invention.

His success convinced the young man that the idea was worthy of a patent. So it was with the help of his mother who had always encouraged him, overcoming the scepticism of his father, that Marconi went to England, where with the help of a cousin on his mother's side, James Davis—who ran a consulting office in London and was very well known among the community of London engineers—he succeeded in attracting interest in his invention from the technical general director of the British Post Office, W Preece, who had recently become a Baron, and in obtaining (2 June 1896) the first patent for the new wireless telegraphy system. Great Britain at the time was at the vanguard in the field of wire telegraphy, both regarding the quantity of traffic and the dominant role of British companies in the construction and exercise of the transcontinental submarine networks, and the choice of Marconi was surely the best one. Between June 1896 and March 1897, Marconi performed a series of experiments on behalf of the Post Office, publicized by conferences made by the same Preece. In 1897 Marconi founded the Wireless Telegraphic and Signal Co Ltd, of which his cousin Davis was the first administrative director, and which in 1900 became the Marconi Wireless Telegraph Co Ltd. On 2 July 1897, a complete patent was issued that described his invention in the most complete and accurate way, elaborated by the consultations of two lawyers, John Cameron Graham, who was an engineer of notable experience in the field of electrical technology, and John Fletcher Moulton, a prince of the London forum, who had studied physics and mathematics at Cambridge, qualifying as senior wrangler, and had specialized in scientific and technological disputes, becoming the major forensic authority in controversies related to the industrial uses of electricity.

Marconi obtained the first wireless transmission (radio) between England and France in March 1899—focusing the interest of the whole world on his experiments—and on 12 December 1901 he realized a link between St Johns, Newfoundland (USA), and Poldhu in Cornwall (Great Britain), on a path of 3200 km, so establishing the first wireless transatlantic telegraphic link. The result was received with stupefaction and great surprise because it was obtained without utilizing the large Poldhu antenna—that was a structure made by 20 pylons 61 m high placed in a circle of 61 m in diameter— nor the station built at Cape Cod in one year of frantic work, because at both stations the antennae had been destroyed by a storm. Therefore at Poldhu, Marconi utilized a much reduced antenna for transmission and a kite for the receiving antenna. The kite was launched not from Cape Cord but from Signal Hill at Newfoundland.

Immediately after this sensational radio-link, O Heaviside (1850–1925) in England, A E Kennelly (1861–1939) in the United States of America, and H Nagaoka (1865–1950) in Japan, in 1902 independently made the hypothesis that reflecting regions for radio waves exist in the high atmosphere. This hypothesis, the only one able to explain why electromagnetic waves may, with straight propagation, overcome the apparent obstacle made by the curvature of the Earth, was experimentally justified by E V Appleton (1892–1965) in 1925, who found that between heights of 100 and 200 km the atmospheric layers are electrically conductive because the gas molecules are ionized by various kinds of agents. Radio waves are reflected towards the earth by the conducting layer so also allowing links between points that have no clear lines of sight. Appleton was awarded the Nobel prize in physics in 1947 'for his investigations of the physics of the upper atmosphere especially for the discovery of the so-called Appleton layer'.

Although in his experiments of 1896 Marconi utilized microwaves (that is waves with a wavelength of the order of a few centimetres) successively he used longer waves and made the first major geographic links with these waves. In fact, Marconi initially believed that electromagnetic waves could reach the receiving antenna only by diffraction, travelling around the Earth's surface very near to it, and therefore made recourse to very long wavelengths. He was convinced that sending signals at increasingly large distances would require waves of longer wavelength, and that to obtain such waves increasingly powerful machines had to be employed. Only in 1916 did he resume the experiments with short waves whose use initially in the 1920s was granted to radio amateurs, because people considered that these waves had no commercial value.

Marconi performed many experiments also in Italy. He had granted the Italian Government the possibility of using his patents *gratis*. So he obtained permission to position an antenna and his apparatus aboard the cruiser *Carlo Alberto*, which from July to September 1902 sailed from Naples to Kronstadt in Russia to accompany King Vittorio Emanuele to the marriage of the son of Tsar Nicholas II. During that cruise he discovered that the distances of transmission increased during the night and decreased in the day: another of the phenomena linked to the presence of a layer of ionized atmosphere around the Earth. With him also embarked Luigi Solari (1873–1957), a lieutenant of the Navy who for the previous few months had been following Marconi's experiments and who worked to smooth the way between the Navy and Marconi and later was the divulger of his life and works. Late in autumn he performed new experiments aboard the same cruiser during the navigation from England to Canada. In 1904 the construction of a high voltage power station at Coltano was started for the Italian Ministry of Mails and Telegraphs. When in 1916, during the First World War, the Italian Navy asked Marconi to study a new device for radio-communications for its navies that was as secure as possible from eavesdroppers, even if of limited range, Marconi returned to short waves, having accepted the determinant role played by ionizing layers. Meanwhile electronic tubes had been invented that gradually replaced spark technology.

The Marconi Company scientific consultant, John Ambrose Fleming (1849–1945), who worked with Maxwell, had been scientific consultant of the Edison Electric Light Company, and had a professorship at London University College, in 1890 invented the thermionic valve, a current-rectifying device based on an effect discovered by Thomas Edison (1847–1931). The Fleming diode consisted of a common incandescent electric lamp in which near the filament, but without touching it, a small plaque was placed and connected to the lamp base by a thin wire. Connecting the plaque to the positive pole of a battery and the filament to its negative pole, a current would flow in the space between the two electrodes. If the polarity is inverted no current would flow. In the first case in fact electrons emitted by the hot filament are attracted by the positively charged plaque, whilst in the second the negatively charged plaque rejects the electrons. The device therefore operated like a valve, from which came the name. In 1904 Fleming recognized the device could have applications and patented it, first in Great Britain and then in the United States (1905). The American inventor Lee De Forest (1873–1961) who took his degree at Yale in 1899 with what probably was the first American thesis on broadcasting, introduced a thin metallic grid between the filament and the plaque in the Fleming diode, building in this way a triode, which he patented in 1907 under the name 'audion'. The grid, if connected in a suitable way, allowed the magnification of the current, and amplified the arriving signal in an extraordinary way. This invention was crucial in the development of wireless transmission systems. The audion was the prototype of the thermionic tube developed in 1912 by Irving Langmuir (1881–1957).

To wage a campaign of experiments to verify the efficiency of short waves, from 1923, Marconi used his yacht *Electra,* which he had bought in 1919 and adapted into a mobile station and laboratory. In 1924 the Marconi Company signed a contract with the English government for the construction of a series of stations that would extend to all colonies of the British Empire and connect Great Britain with Australia, India, South Africa and Canada. The company decided to use short waves. The first 'bridge' of the network was inaugurated in 1926.

The Marconi decision to use short waves for these stations was a radical change of technology. Until then, the Marconi Company had encouraged the use of ever longer wavelengths. Now, Marconi, on the basis of experiments he did for the Italian Navy, in which he had found that short waves could propagate for much longer distances than foreseen, made this decision, accepting the thesis of reflection from the ionized layers of high atmosphere, understanding that it was time to give up the blind alley of long waves that could no longer be continued. In 1923 a transmitting station at short waves was installed by the Marconi Company at Poldhu utilizing thermionic tubes, and Marconi started his experiments with the *Electra.*

In 1928 he was appointed the president of the Italian National Research Council (CNR) and in 1932 installed a system of short wave radio-telephony between the Vatican and Castel Gandolfo, the papal summer residence near Roma.

For his activity Marconi was awarded the physics Nobel prize in 1909 together with the German physicist K F Braun (1850–1918) who besides having invented a crystal diode and the oscilloscope had improved wireless communication systems with the introduction of suitable circuits. Marconi was the president of the Italian Academy, a friend of Mussolini and received the title of Marchese. When he died he received a State funeral. All the wireless telegraph and telephone offices in the British Isles observed two minutes of silence during his burial.

Popov

In Russia, the employment of radio waves for communications was developed independently from Marconi, by Professor Aleksandr S Popov (1859–1906) who developed one of the first receivers of electromagnetic waves. Augusto Righi wrote 'the new characteristics of the Popov apparatus are the use of a hammer electrically moved or electric bell, to reconstruct the initial electrical resistance of a coherer, and the use of a vertical conductor, later called antenna, to detect the waves'.

Aleksandr Popov was born (1859) in an industrial village of the north Urals, the son of a priest, and was supposed to follow an ecclesiastic career according to the family tradition. His passion for science, instead, made him frequent the Physics and Mathematics Department of the St Petersburg University where he brilliantly received his degree with a thesis on electrical machines. In 1883 he was asked to teach at the Torpedo School for officials of the Russian Navy at Kronstadt. The School had been established in 1874 and was the most advanced Russian institution in the field of electric techniques. Popov spent 18 years at the School, interesting himself in the physics laboratory and performing several teaching courses. He became an authority in the field of electricity, being asked many times by the Russian Navy for the solution of problems of practical interest.

After his successes, in 1901, he was appointed Professor at the Electrical Engineering Institute in St Petersburg, and in 1905 was elected rector. Early in the 20th century the tensions between Russia and Japan had become worse, and in 1904 the Russian–Japanese war exploded. 1905 was characterized by tumultuous political events, strikes and demonstrations throughout the whole country. The government in the December of that year imposed among other things a prohibition on holding public meetings in the rooms of the Institute. Popov refused to accept the order, and becoming ill due to the stress to which he was subjected between students and authorities, died suddenly of a cerebral haemorrhage in January 1906.

After Hertz's publications in 1888–89, Popov became interested in Hertzian waves and, having known about the coherer, early in 1895 performed a series of studies to make it as reliable as possible by inserting a small hammer that was triggered when current flowed in the device and with a small knock restored the initial conditions (figure 26).

The first demonstration of the receiver was given at the Physics Society of St Petersburg on 7 May 1895 and Popov exhibited the transmission of intelligible signals in March 1896 at a meeting of the Society. At the time he was teaching at the Navy School in Kronstadt and his results were not published because they were classified as military secrets.

By performing experiments in the open air he observed, as already Lodge and others had done, that the coherer reacted to atmospheric disturbances and its sensitivity could be increased when one of its extremities was connected to a vertical wire attached to a balloon or to a lightning-rod, while the other extremity was connected to the ground. Popov capitalized on this to

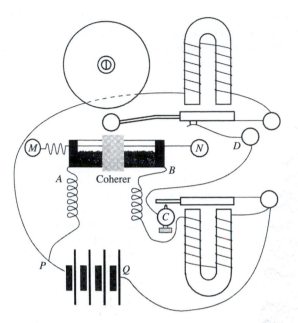

Figure 26. The Popov system for the detection of electric oscillations. The layout shows the electric connections and the positions of the parts. (From the paper by A S Popov 1896 'Apparatus for the detection and recording of electrical oscillations'. *J. Physico-Chemical Russian Society Physical Section* **1** 1–14 in Russian).

build a lightning detector that was erected at the Forest Institute of St Petersburg. With it he also performed experiments on transmission in the laboratory and gave public demonstrations in 1896, and he installed another detector at the electric plant of the Fair of Industry and Arts of the Russian Empire, which was held every year in the summer at Niznij Novgorod on the Volga. The plant for which he was responsible was built in 1885, the most important in Russia for the quantity of produced energy (up to 400 kW) and for the dimensions of the distribution network. The lightning detector indicated the approach of a storm and allowed the network to be switched off to prevent damage to the plant. While at the 1896 Fair, Popov read of Marconi's experiments, and with the encouragement of the Russian Navy resumed his transmission experiments, continuing to investigate the propagation of electromagnetic waves and working to develop the first wireless communication systems for the Navy. However, he was obliged to rely on foreign manufacturing for the production of these systems because in Russia there was no adequate industry. The Parisian engineer and businessman Eugene Ducretet (1844–1915), who was the first in France to build wireless telegraphy devices, was very interested in Popov's results and, in 1898, started to build radiotelegraphic stations based on his system. The Ducretet–Popov collaboration was fostered by the political re-engagement between Russia and France that occurred at the end of the 19th century. Between 1899 and 1904 the Ducretet Company obtained several orders from the Russian Navy. However, the Company was too small and during the Russian–Japanese war of 1904–05, the Russian Navy was obliged to use broadcast systems made in Germany by Telefunken.

Microwaves

Microwaves, as we have already said, were the last to be utilized, even though they were the first to be produced. To obtain them it was necessary to decrease both the dimensions of the tubes, then used for oscillating circuits, and the dimensions of the circuits. Soon a problem surfaced caused by the time required by electrons to move from the grid to the plaque in the tubes.

We recall that in a vacuum tube, such as those used then, electrons are emitted by a filament heated by a current in an evacuated glass bulb, and crossing a metallic grid, they are collected by an electrode named the anode, generating in this way a current. This current may be controlled by the electric potential given to the grid. Obviously to move from the filament to the anode, travelling through the grid (all these elements are indicated with the name of 'electrodes'), electrons take a certain time and if during this time the electric potential of the grid changes in an appreciable way, the received signal at the anode is distorted.

To avoid this inconvenience people tried to build tubes smaller and more compact in which the distances between filament, grid and anode were reduced to a minimum. The problems connected in the use of conventional tubes were described very well in a paper by Irving Langmuir and Karl Compton in 1931 in the United States, in which they indicated how the high frequency limit could be extended, simply by reducing the dimensions and distances between electrodes.

Already in 1933 in the States, the Radio Corporation of America (RCA) had built the arcon tube and the Western Electric the famous 'door-knobs'. These miniature tubes were able to produce oscillations up to about 1500 MHz (that is wavelengths of about 20 cm). However, they produced only very low power.

The magnetron

Thus a new device, the magnetron, entered the stage and became the preferred generator around the middle of the 1920s, and it was discovered it could work in such a way as to produce oscillations at very high frequencies.

The magnetron combines the use of an electric and a magnetic field, and in the first realizations it consisted of a filament of straight wire, surrounded by a cylindrical anode. A magnetic field was oriented in a direction to influence the motion of the electrons in order to make them move in a spiral between the two electrodes.

The new device had been invented by Albert W Hull (1880–1966) who was born in a Connecticut farm and after having earned his degree in physics at Yale and taught for some time, in 1913 entered the research laboratory of General Electric (GE). In 1914 he invented the 'dynatron', the first of a long series of tubes he pioneered. He also conducted research on crystallography problems and the use of x-rays.

During 1916, Hull started experiments on the control of the electron flow in tubes by means of magnetic fields, as an alternative way to the electrostatic control by means of the grid as was then in use. The control by means of a grid was at the time the object of a controversy between GE and the American Telephone and Telegraphy Company that had bought the royalties of the invention from its author, the American Lee de Forest.

In the 1920s, Hull and his collaborators at GE demonstrated that the device, having acquired several names but since 1920 assuming the title of 'magnetron', could be used at low frequencies as an amplifier or an oscillator in radio systems or as an electronic switch in power converters. In the summer of 1921, an agreement on the rights for the production of tubes made the magnetron seem less important for radio applications. GE continued its development, however, for high power applications, and in 1925 Hull had built a high power magnetron that produced waves of 15 km wavelength with a power of 15 kW.

The important discovery that a magnetron could generate oscillations at very high frequency was made, independently in Germany and Japan, in the middle of the 1920s but was not generally known in America until 1928. The Czech physicist August Zacek published in Czech, in 1924, the results of experiments in which he succeeded in producing oscillations at 29 cm. However, only when his work was reproduced in a German specialized journal in 1928 did it receive significant diffusion. Again in 1924 similar results were obtained by Erich Habann of Jena University.

In Japan the electrophysicist Hidetsugu Yagi (1886–1976) and his student Kinjiro Okabe (1896–1984) made important contributions to the development of the magnetron at high frequencies. Yagi was born in Osaka and took his degree in engineering at Tokyo University in 1909. Before the First World War he studied in England with John A Fleming, the inventor of the valve diode used to detect radio waves. Yagi became interested in the possibilities of communications with short waves when he spent some time in Dresden, in Germany, with Heinrich Georg Barkhausen (1881–1956), the inventor of a particular tube oscillating circuit for high frequencies. At the beginning of the war, Yagi came back to Japan to teach at Tohoku Imperial University in Sendai, where he earned his doctorate in 1919. In the early 1920s he learnt of the Hull magnetron from a Japanese Navy official who was returning from a visit to the United States.

Okabe, who became a first rank researcher on magnetrons, obtained his degree in engineering from Tohoku University in 1922 and earned his doctorate in 1928 under the supervision of Yagi. In 1927 Okabe announced he had succeeded in producing oscillations at a wavelength of about 60 cm using a magnetron. He had investigated a great variety of different electrode geometries and had found that by dividing the cylindrical anode into two semi-cylinders (a configuration known as 'split-anode') a greater power was obtained. In 1928 Yagi paid a visit to the United States to discuss the Japanese results and presented the last experiments of Okabe who had succeeded in the meantime in producing wavelengths of 12 cm. He described also a directional antenna for very high frequencies that he had developed which consisted of an active element and several reflecting and directing passive elements; this is the antenna that we still use as the external receiving antenna for television.

After a link realized in 1931 by a French–English group across the English Channel, using waves of 18 cm, great interest arose throughout the world for microwave communications. A journal published an editorial stating the system had opened a true no man's land that could provide a range of frequencies for thousands of radio channels. The director wrote that the experiment through the English Channel marked a 'new epoch in the art of electrical communications' so revolutionary as to demand a new name. He remarked also that the apparatus was so compact that similar systems could be installed on ships or used in aeroplanes.

Also the very well known American specialist periodical *Electronics* announced the amazing link between Dover and Calais and in October two years later observed that ultrashort wavelengths, once considered of no use, had suddenly acquired great importance.

The magnetron began its rise to fame: the number of scientific publications about this device increased clamorously in 1933 and remained at a high level until 1940 when, with the outbreak of the war, the publication of scientific results stopped for security reasons. Important research that resulted in many significant improvements were made in the 1930s in France, England and Germany.

The klystron
In the 1930s only one good oscillator for very high frequencies was available: the magnetron. C E Cleeton (1907–) and N H Williams (1870–1956) of the Michigan University performed the first spectroscopic measurements at microwave frequencies at that time, using a magnetron to investigate the absorption spectrum of gaseous ammonia. In research to determine the practical limit of the wavelengths that could be generated by magnetrons, in 1936 they announced they had generated oscillations at 6.4 mm. However, the magnetron efficiency was not very high. As such, a new device, named the

'klystron', was developed at Stanford in California by Russel Varian (1898–1959), Sigurd Varian (1901–1961), David Webster, William Hansen (1909–1949) and John Woodyard.

The klystron operated on a completely different principle from that of any other system used to produce oscillations at ultrahigh frequencies. In it electrons are forced to bunch into packets and successively supply energy to a resonant cavity.

A resonant cavity is made by a hollow conductor in which standing electromagnetic waves are excited. To make the cavity resonate, the wavelength of radiation needs to have some relationship with the cavity dimensions. For cavities of a simple shape, for example a cube, this relation says that the side of the cavity must be an integer multiple of half a wavelength, as we already have seen. By decreasing the wavelength, the microwave production technique evolved. It was found that to transmit microwaves from one side to the other in a circuit it was sufficient to guide them into suitable metallic structures. The waveguides, as they are called, are metallic tubes of cylindrical or rectangular section and waves are propagated by successive reflections at their walls. These waveguides may be used also as antennae if their terminal end is open.

Radar

Microwave instrumentation and production techniques received a vigorous impulse between 1930 and 1945 due to the need to produce ultrashort waves for the development and construction of radar (the acronym was introduced by Americans for radio detection and ranging).

The operating principle of radar is very simple: a pulse of radio waves is sent towards a target which partially reflects it back towards the receiver where it is collected. The sent and received pulses are visualized on an oscilloscope and by measuring the time interval between the moment at which the pulse has been sent and the time at which it comes back, the distance to the target can be obtained.

Already Hertz, among others, had shown that radio waves are reflected by solid objects. In 1904, the German engineer C Hulsmeyer possessed a patent to use this property to detect obstacles and aid the navigation of boats. He built a device with which he obtained good results in the Rotterdam haven, but nobody was interested in developing the system, which was too advanced with respect to other ideas of the time.

The results of the first experiments of ionospheric radio-probing, performed by Appleton (1925) to elucidate the existence of ionizing layers in the atmosphere capable of reflecting microwaves, suggested afresh the idea of using methods based on the reflection of radio waves for the localization of objects situated at large distances. The principle of using radiation pulses that characterizes modern radar, was used for the first time in 1925 by G Breit

(1899–1981) and M A Tuve (1901–1982) at Washington Carnegie Institute to measure the height of the ionosphere. So a number of experiments to apply the principle for the localization of earth-bound objects and for the measurement of their distance started in Europe and the United States. During research into using microwaves for communication it was found that the presence of boats and planes could be detected by the energy they reflected.

In America, at the Naval Research Laboratory, the possibility of detecting moving objects by the reflection of electromagnetic waves was already well known in the 1920s, and W Delmar Hershberger and his collaborators at the Army Signal Corps Laboratory at Fort Monmouth, New Jersey, employed magnetrons built by Westinghouse and RCA for detection experiments with centimetres waves, starting from 1933. Hershberger had worked at the Naval Research Laboratory during 1927–31 and was familiar with the research effort on the detection of radio waves. Westinghouse had exposed a microwave system that used a magnetron able to produce 9 cm waves at the Chicago World Exhibition in 1933. A little later, researchers from Fort Monmouth bought two of the Westinghouse magnetron transmitters for communication experiments and observed reflection signals from a moving truck near the transmitter. Successively, in the summer of 1934, in collaboration with RCA engineers, they detected reflection signals from a ship at the distance of half a mile. In 1936 the group at Signal Corps concluded that the available magnetrons did not produce enough power for practical detecting systems on aircraft and diverted their effort to longer wavelengths, for which they could use conventional tubes to produce relatively higher powers. However, at this time there was a lack of specialist researchers in this area for a working device to be realized.

Radar development in Great Britain

In Great Britain in 1934, the desire to protect the country from air attack led H E Wimperis, the Director of Research in the Air Ministry to seek the advice of A V Hill (1886–1977), the distinguished Cambridge physiologist, Nobel prize winner for physiology in 1922, and a gunnery officer in the First World War, about possible means of destroying enemy aircraft. The result of these conversations was that on 12 November 1934 Wimperis sent a memorandum to Lord Londonderry, who was then Secretary of State for Air, asking him to set up a committee to consider how far recent advances in scientific and technical knowledge could be used to strengthen the methods of defence against hostile aircraft. Wimperis suggested that the Chairman of the Aeronautical Research Committee, Professor H T Tizard (1885–1959), a Dean of chemistry at Imperial College, should be the chairman of the committee and that the other members should be A V Hill and Professor P M S Blackett (1897–1974) 'who was a Naval Officer before and during the War (1914–18), and has since proved himself by his work at Cambridge as one of the best of the younger scientific leaders of the day', who did not disprove his potential being awarded the physics Nobel prize in 1948 'for his development of the Wilson cloud chamber method, and his discoveries therewith in the fields of nuclear physics and cosmic radiation', and A P Rowe, an assistant to Wimperis, acting as secretary. The committee was immediately settled on and on 28 January 1935 the first meeting was held. As one of the first actions, Wimperis approached the Superintendent of the Radio Research Station of the National Physics Laboratory, Robert Watson-Watt (1892–1973), who would receive a Knighthood in 1942, to inquire whether or not it would be possible to incapacitate an enemy aircraft or its crew by means of an intense beam of radio waves. Watson-Watt replied immediately in a memorandum that to produce such a 'death-ray' was not practicable; it was instead possible to detect the aircraft by radio waves. Upon request, in February 1935 Watson-Watt produced in a second memorandum a series of calculations, showing that the energy reflected by an aircraft illuminated by a powerful radio beam could be used for that purpose. The memorandum was submitted to Tizard's committee. A demonstration was immediately commissioned and gave excellent results on 26 February 1935. The demonstration was so successful that on 13 May a team of five people (A F Wilkins, L H Bainbridge-Bell, E G Bowen, G Willis and J Airey) were sent from the Radio Research Station at Slough to Orfordness to begin the development of a Radiolocation and Direction Finding (RDF) system as it was called by the British at the time. Five weeks later they detected a Scapa flying boat at a range of about 30 km and showed it was possible to detect and plot the position of aircraft at a range of 160 km. Then they established a new headquarters a few miles away at Bawdsey Manor.

Robert Watson-Watt had written the fundamental equation for radar which shows that, for an aircraft, the maximum detecting range is proportional to the linear dimension of the antenna and to only the fourth root of the mean power. The fourth root means that to double the distance one needs to use a power 16 times higher. The distance increases also by decreasing the wavelength, but this was not of interest to Watson-Watt. At this time there were no transmitters of high power for wavelengths below 10 m. Therefore a wavelength of 50 m was initially chosen on the assumption that bomber aircraft would reflect best when their wing-span was roughly equal to half a wavelength. But it had soon been found that there was far too much interference from radio stations at great distances and the wavelength was reduced to 26 m. Experience showed, however, that for the chain of radar stations which it was proposed to build round the coast, an even shorter wavelength would be needed. At Orfordness, work was begun using a wavelength of 13 m.

By September 1938 the Thames estuary was covered by a radar chain for detecting aircraft and radar equipment was installed on all the major

British battleships. Because of this preparation, Great Britain could resist and turn to its favour the air battle waged against it by Germany during the Second World War (August 1940) while the installation of radar on ships allowed the English navy to obtain notable successes (Battle of Cape Matapan, March 1941).

In March 1935 the only device on ships of the Royal English Navy to detect aircraft was a binocular with a magnification of 7. Shortly after the Munich crisis (1938) they could detect an aircraft 100 km away thanks to radar.

The development of naval radar continued during the war, partly along a somewhat strange path under the pressure of war-related events. One episode may help us understand how the use of radio waves was at the time poorly mastered and some effects appeared mysterious. In 1943 the Germans started attacks on the English navy using bombs released by planes and controlled via radio, creating great anxiety among the crew. A naval scientist by chance one day switched on his electric razor during one of these attacks in the Bay of Biscay, and to his great astonishment and that of the crew, the bomb started to turn in the sky and eventually hunted the aircraft that launched it. Immediately the Admiralty sent the order to switch on, under attack, all the available electric razors and agitate them against the missiles. The efficiency of this countermeasure is not documented in the official papers, but it certainly bolstered the morale of the men.

After the development of the radar chain to protect the Thames, Watson-Watt and his collaborators turned their attention to the radar to be installed on aircraft. A second generation of radar at the wavelength of 1.5 m could have sufficiently small dimensions to be installed on a plane to transform it into a night fighter and an identifier of submarines. Everybody agreed that radar at the wavelength of 10 cm would be much better and it was in trying to improve the range of wavelength used that on 12 August 1940 the first radar echoes were obtained from aeroplanes using wavelengths of 10 cm.

One of the reasons to use shorter wavelengths is that the dimensions of the details that can be revealed are proportional to the wavelength adopted and therefore, by decreasing it, the definition of the target image increases.

The cavity magnetron
In the first months of the war the British had formed a Committee for the Coordination of Valve Development that secured a number of industrial and research contracts for the development of transmitting and receiving valves at a wavelength of 10 cm. One contract was placed with the group of Professor M L E Oliphant (1901–2000) at Birmingham University. Oliphant was a physicist who in 1937 had moved from the Cavendish Laboratory to set up a nuclear physics laboratory at Birmingham. The group was initially interested in developing an oscillator with a klystron

133

that Oliphant had seen during a visit to the United States in 1938. So, by the end of 1939 a klystron was assembled which produced about 400 W of continuous power at a wavelength of 10 cm and in a few months another one was produced which could be pulsed and used for radar.

However, the device was too cumbersome to be fitted in airborne equipment and although smaller klystrons were used in airborne radar, an alternative solution was sought. In the group also worked J T Randall (1905–1984) who earned a doctorate from Manchester University and had performed research on x-ray scattering, spending a year at the British General Electric Company Laboratory at Wembley, where he acquired practical experience with vacuum tube technique, and H A H Boot (1917–1983). Boot had an undergraduate degree in physics from 1938, and later obtained a doctorate at Birmingham in 1941. Oliphant had asked them to investigate a special circuit which required a powerful oscillator to operate and therefore in 1940 they started studying how to improve a magnetron and had the idea of using a structure with a cylindrical resonant cavity that served the dual function of determining the frequency and serving as the magnetron's anode. They tried the new device that was named a 'cavity magnetron' on 21 February 1940 and achieved a continuous power level of about 400 W at 9.8 cm.

Immediately, the device was produced by GE at Wembley with some important modifications increasing its power to 10 kW, and by May a radar using the new cavity magnetron was already working and able to detect a submarine periscope at a distance of 10 km.

In the autumn of 1940, a British technical and scientific mission, headed by Sir Henry Tizard, brought the cavity magnetron to the United States. After the triumph of the Thames defence, one may have expected that Tizard was chosen as the principal Defence scientific consultant. For some time it was so, but when Winston Churchill became the prime minister, he chose instead the physicist Frederick Lindemann (1886–1957), the director of the Clarendon Laboratory at Oxford. Tizard lost much of his power but was, however, charged with the delicate mission in the States. The mission should have pushed the United States, at this time neutral, to develop and build the new technology necessary for the new war. The British at first hesitated to divulge the design of the magnetron to the Americans for fear that it would fall into the hands of German intelligence, but the subsequent developments completely justified the Tizard commission's actions. So on 6 October 1940 the cavity magnetron was demonstrated for the first time to Americans at Bell Telephone Laboratories. The disclosure of this device led to the formation later that year of the Radiation Laboratory at Massachusetts Institute of Technology, the famous MIT at Boston. An elite group of scientists and engineers recruited from universities and industry developed there a variety of magnetrons during the war years and incorporated them into more than 100 radar systems, giving the Allied forces a decisive technical advantage. More than two billion dollars were invested in the development and production of radar systems by the United States during the war, and a momentum in microwave technology was created that persisted in the post-war period.

On the suggestion of Tizard, many of the personnel recruited were nuclear physicists because, as the British visitors said, through their experience, they would adapt more easily to the new research than radio engineers. Lee DuBridge (1902–1994) of Rochester University was appointed the director of the new laboratory. In it were called up, among others, I I Rabi, Ken Bainbridge (1904–1996), Norman Ramsey, Ed Purcell, Ernie Pollard and Luis W Alvarez, many of whom will play a role in our story.

Radar in other countries

During the conflict, radar was developed for air defence purposes also by the Germans and for navigation purposes by the Italians (1941). The Germans had already started research on magnetrons and microwave systems in the 1930s. Telefunken in Berlin experimented with a system possessing a 'mysterious ray' to detect aircraft that was described in a paper published in 1935 in the American journal *Electronics*. The firm developed the *Wirzburg* which was the Luftwaffe's principal anti-aircraft radar throughout the war.

It seems, however, that the proposal to equip fighter aeroplanes with radar encountered the opposition of Hermann Goering who maintained that the ability of German pilots was sufficient not to require a 'cinematographic' instrumentation. The firm GEMA (Gesellschaft für Elektroakustische Mechanische Apparate) built the first functioning radar set in 1935. In 1938 Hitler, Goering and chief officials assisted to the demonstration of the GEMA air-warning radar. GEMA developed first for the German Navy and later for the Luftwaffe several radar sets among which the famous Freya air-warning and Seetakt ocean-surveillance radars. The rivalry of the German forces, the lack of involvement of universities and the limits imposed by severe secrecy were certainly among the causes of difficulties encountered in fully exploiting radar in Germany.

In Italy in 1924, Professor Nello Carrara (1900–1993) was appointed to the physics chair at the Naval Academy in Livorno in which was lodged the Royal Electro-Technical and Communications Institute directed by Admiral Professor Giancarlo Vallauri (1882–1957). Here he continued his research on the generation and detection of centimetre waves. In 1931 he succeeded in building systems which formed a link between Livorno and La Spezia with waves at 30 cm and around that time he coined the term 'microwaves'. Early in the 1930s, various authors wrote about the possibility offered by microwaves to produce an echo from fixed or mobile obstacles, and in 1933 Marconi performed a successful experiment using motor vehicles moving near a microwave beam. The news was distorted by the media who spoke of a 'death ray' with which Marconi had succeeded in stopping the engines of cars and planes.

In 1935, Professor Tiberio (1904–1980), an officer of the Navy, presented to a Ministry commission a lecture in which he demonstrated the feasibility of using microwaves for night detection (so it was written at that time), and the following year the Navy started secret military research at the Livorno Institute. Tiberio guided the research with the objective of developing a radio-telemeter. In the project also participated Carrara who developed systems. From 1936 to 1941 several prototypes were realized and, early in 1942, 50 systems for the Navy and some for the detection of aircraft from the ground were assembled. With one of these system mounted on a high terrace of the Naval Academy, the American air formations that in May 1943 performed the first violent bombing of Livorno were detected. In this construction phase, great problems were encountered in the supplying of the necessary components because the goods from United States that were the customary suppliers were unavailable and Germany demanded unacceptable conditions. The decision was therefore taken to produce the components in Italy and the task to establish the national industry was given to Carrara. A great effort was made regarding the production of prototypes, but, unfortunately, industrial support was missing owing to the time required to organize it. It was so that, in the Battle of Matapan, Italian ships were deprived of radar, with the known result.

In Japan, notwithstanding an intense research effort into magnetrons and microwaves, the development of radar systems was delayed by the lack of cooperation between the Army and the Navy and the lack of a centralized effort comparable with that made in America at MIT. In the Soviet Union, research on magnetrons was active but apparently the decisive step of using pulses rather than continuous waves was not taken.

During the war many of the best English and American scientists were involved with research on microwaves and radar, and at the end of the war microwave systems were easily passed on to research institutes to enable research in this field to be continued. At the end of the war part of this research was therefore in a very natural way principally centred on problems connected to microwaves.

The interaction of microwaves with matter may produce transitions between energetic levels of molecules that, being very near one to the other, correspond to the low energy carried by the microwave photons. Microwaves may also interact with the magnetic moment of the electron (spin) or of the nuclei. In these cases the magnetic field of the waves interacts with the magnetic moment of the particle (electron or nucleus) and orients it by exchanging energy. These phenomena of interaction between microwaves and matter are the object of microwave spectroscopy. Microwave spectroscopy was born naturally from the development, during the Second World War, of radar and microwave oscillators. After the war these oscillators were used for spectroscopic studies that allowed the smaller details of the molecular structure and of the atomic nuclei to be measured. The only measurement performed before the war with microwaves was that of the inversion frequency of ammonia in the centimetre region performed by Cleeton and Williams in 1934, of which we have already spoken.

Due to its scientific interest, the work passed rapidly from the industrial laboratories, in which applied research was carried out, to the universities where information concerning fundamental questions of physics and chemistry were sought. For these studies the frequency purity or coherence of the radiation was a very important characteristic. It was necessary to have sources that emitted a single frequency or, if this was not possible, at least in a very limited range of frequencies without suffering too many fluctuations.

To better understand the interaction between microwaves and matter we need, however, to come back to the further developments of spectroscopy.

Plate 1. The birthplace of Newton, Wollsthorpe Manor in England. (Photograph by Roy L Bishop, Acadia University, with kind permission of AIP Emilio Segrè Visual Archives.)

Plate 2. Joseph John Thomson. (AIP Emilio Segrè Visual Archives.)

Plate 3.   Johann Jakob Balmer. (AIP Emilio Segrè Visual Archives, W F Meggers Collection.)

Plate 4.   Wilhelm Wien. (Photograph by Ullstein, with kind permission of AIP Emilio Segrè Visual Archives.)

Plate 5.   Gustav Robert Kirchhoff. (AIP Emilio Segrè Visual Archives, W F Meggers Collection.)

Plate 6.   Ernest Rutherford (right) with J A Ratcliffe at the Cavendish Laboratory, Cambridge, in 1935. (Photograph by D Schoenberg, with kind permission of AIP Emilio Segrè Visual Archives.)

Plate 7. Niels Bohr (left) and Max Planck (right) in 1930. (AIP Emilio Segrè Visual Archives, Margrethe Bohr Collection.)

Plate 8. Niels Bohr (left) and Wolfgang Pauli (right) in Brussels for the Solvay Congress. (CERN, with kind permission of AIP Emilio Segrè Visual Archives.)

Plate 9. Albert Einstein (right) receives a medal from Max Planck (left) in 1929. (**AIP** Emilio Segrè Visual Archives, Fritz Reiche Collection.)

Plate 10. Participants at the first Solvay Congress in Brussels, 1911. (Institut International de Physique Solvay, with kind permission of AIP Emilio Segrè Visual Archives.)

Plate 11. From left to right: Walther Nernst, Albert Einstein, Max Planck, Robert Andrews Millikan and Max von Laue in Berlin, 1928. (AIP Emilio Segrè Visual Archives.)

Plate 12. Robert Andrews Millikan (left) and Otto Stern (right) in 1928. (AIP Emilio Segrè Visual Archives, Segrè Collection.)

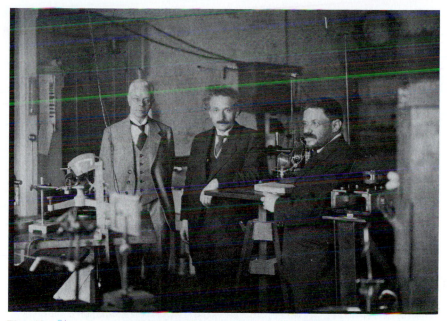

Plate 13.   Pieter Zeeman (left), Albert Einstein (centre) and Paul Ehrenfest (right) in Zeeman's laboratory in Amsterdam. (AIP Emilio Segrè Visual Archives, W F Meggers Collection.)

Plate 14.   Hendrik Antonie Kramers in Leida, 1937. (AIP Emilio Segrè Visual Archives, Lindsay Collection.)

Plate 15.   Rudolf Walther Ladenburg in his laboratory. (AIP Emilio Segrè Visual Archives.)

Plate 16.   Einstein and Ladenburg on the occasion of Ladenburg's retirement from Princeton University. (AIP Emilio Segrè Visual Archives.)

Plate 17. James Franck (left) and Gustav Ludwig Hertz (right) at the railway station in Tubinga. (AIP Emilio Segrè Visual Archives.)

Plate 18. Heinrich Hertz. (Deutsches Museum, with kind permission of AIP Emilio Segrè Visual Archives.)

Plate 19. Isidor Isaac Rabi in his laboratory. (AIP Emilio Segrè Visual Archives.)

Plate 20. Felix Bloch in 1973. (AIP Emilio Segrè Visual Archives, Segrè Collection.)

Plate 21.   Edward Mills Purcell. (AIP Emilio Segrè Visual Archives, Segrè Collection.)

Plate 22.   Polykarp Kusch. (AIP Emilio Segrè Visual Archives, Physics Today Collection.)

Plate 23.   Norman Foster Ramsey in 1952 with the molecular beam equipment at Harvard. (AIP Emilio Segrè Visual Archives, Ramsey Collection.)

Plate 24. Nikolaj Gennadievic Basov (left), Charles Hard Townes (centre) and Aleksandr Michailovic Prokhorov (right) in the Soviet Union, 1965. (Sources for the History of Lasers, with kind permission of AIP Emilio Segrè Visual Archives.)

Plate 25. Charles Hard Townes (left) and James Gordon (right) in c. 1954 with the second maser at Columbia University. (AIP Emilio Segrè Visual Archives, Physics Today Collection.)

Plate 26. Nicolaas Bloembergen.
(AIP Emilio Segrè Visual Archives.)

Plate 27. Theodore Harold Maiman
in c. 1967. (AIP Emilio Segrè Visual
Archives, Physics Today Collection.)

Plate 28. Arno Penzias (left) and Robert Woodrow Wilson (right) in 1978. In the background is the radio-telescope with the large horn-shaped antenna used to discover the background radiation. (AIP Emilio Segrè Visual Archives, Physics Today Collection.)

Plate 29. Robert Henry Dicke in 1962. (AIP Emilio Segrè Visual Archives.)

Plate 30. Arthur L Schawlow in c. 1963 with the first ruby laser at Stanford University. (Stanford University, with kind permission of AIP Emilio Segrè Visual Archives.)

CHAPTER 8

SPECTROSCOPY, ACT II

After the seminal work of Bohr in 1913, atoms and molecules were at the centre of interest for both theoretical and experimental physicists, while at the same time knowledge of the atomic nucleus was increasing through studies on radioactivity and nuclear reactions obtained by bombarding heavier atomic nuclei with nuclei of hydrogen, helium and other particles.

If, before Bohr, spectroscopy was an essentially empirical science that went little farther than to catalogue absorption and emission wavelengths, the new atomic theory offered a guide to the interpretation of the experimental results and, as often happens, theory and experiment cooperated with each other to yield the interpretation of various observed phenomena.

The first proof of the existence of discrete atomic energy states
The centre of gravity for research had now moved to Germany. One of the most noteworthy results was obtained in 1913–14, before the war, by James Franck (1882–1964) and Gustav Hertz (1887–1975).

> Franck, the son of a Hamburg banker, was educated in Heidelberg and Berlin and obtained the experimental physics chair in Göttingen, which he left in 1933 after Hitler rose to power, emigrating to the United States and joining Chicago University. During the Second World War he participated in research on the atomic bomb, adopting thereafter a stance against its use.
>
> Gustav Hertz, also from Hamburg, the nephew of Heinrich Hertz, was severely wounded during the war, and when he returned to Berlin in 1917 the only available position was an unpaid lectureship at the University. In the early 1920s he joined the newly founded research laboratory of the Philips Company in Holland, one of the very first industrial laboratories in which basic research was pursued. In 1925 a partnership in Germany at the University in Halle was finally offered to him and then he became professor of experimental physics in Berlin from 1928 to 1935, when he was obliged to resign because he was Jewish. He then worked for the German company Siemens from 1935 to 1945, surviving the war, and was then captured by the Russians. In 1955 he reappeared as the director of the Physics Institute in Leipzig in the former East Germany.

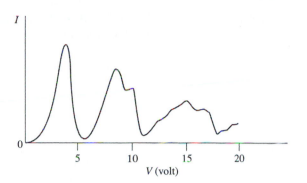

Figure 27. The result of the Franck and Hertz experiment for mercury vapour. On the ordinates the current as a function of the applied potential is shown. From the values of the potential for which current minima occur, the excitation energies of electrons in the atom can be derived.

The two physicists had designed an ingenious experiment to measure what they thought was the ionization energy of gaseous atoms, that is the quantity of energy that must be given to an atom to liberate its outermost electron. The experiment consisted of exciting atoms through impact with electrons, and then, by measuring the energy lost by the electrons in the collision, the exchanged energy was calculated. They found that a series of energy exchanges was possible as shown in figure 27 where each current minimum, measured by increasing the electron energy, corresponds to an energy transfer from electrons to atoms. Bohr gave the correct interpretation of the values that they had determined for the 'ionization energies' by reinterpreting them as those corresponding to the energies of the various possible orbits for electrons in the atom, and was able to extract from their findings not only the conclusion that an atom does not exchange energy with an electron whose energy is insufficient to 'ionize' it, but also that the orbits of electrons in an atom can have only certain well-defined energies. The Franck and Hertz experiments and the Bohr interpretation demonstrated in a conclusive way not only the existence of the stationary states, postulated by Bohr, but also that these states could be excited with electronic collisions and that the jumps among them obey the fundamental law postulated by Bohr.

For their work Franck and Hertz were awarded the physics Nobel prize in 1925.

The development of Bohr's theory

In spite of these results, and although the declared purpose of Bohr's work of 1913 was to develop a general theory of the constitution of atoms and molecules, this theory gave a rigorous and adequate explanation only of the hydrogen atom and of the hydrogen-like atoms (that is of those atoms from which all electrons have been stripped save one), and all efforts to extend it to systems with more than one electron failed. Even the spectrum of neutral helium, which as we said is formed of a nucleus around which only two electrons turn, could not be explained.

One of the successes that Bohr's theory achieved was the explanation of the line series observed by the American astronomer W H Pickering (1858–1938) in the spectrum of the star ζ-Puppis. They were assumed to be hydrogen lines, because they had many similarities with the Balmer series, but Bohr showed they actually were produced by ionized helium in which a single electron is linked to a nucleus with charge $+2$. Einstein was at a conference in Vienna in September 1913 and when he was informed of the result exclaimed 'Then the frequency of the light does not depend at all on the frequency of the electron [that is on how many turns the electron completes around the nucleus per unit time]. And this is an enormous achievement. The theory of Bohr must be right.'

In his studies of 1913, immediately after discussing the theory of the hydrogen atom, Bohr turned to atoms containing several electrons. He represented these atoms as systems consisting of a positively charged nucleus surrounded by electrons moving in circular orbits and wrote: 'We shall assume that the electrons are arranged...in coaxial rings rotating round the nucleus'. The problem was to know how many electrons could reside in each ring with the configuration remaining stable notwithstanding the repulsive electric forces existing between electrons. Bohr tried to solve the problem by applying classical dynamics and started to derive the configurations of the simplest atoms: for helium, which has two electrons, he assumed correctly that they were bound in the same orbit, for lithium (three electrons) he assumed, still correctly, that two electrons reside on the innermost orbit (ring) and the third one in a larger orbit (a new ring), and for beryllium, that has four electrons, he assumed that two electrons are on one ring and two on another ring (also this hypothesis was later found to be correct). For atoms with a larger number of electrons the considerations became very cumbersome. In the end he arbitrarily established that the number of electrons that resided on the outer rings was equal to the number that chemists gave for the chemical valence of the element.

The atom was assumed to be flat, that is the nucleus and all electrons were assumed to lie in a single plane. Bohr's hypotheses were plausible but insufficient to fix unequivocally the distribution of electrons among the various rings around the nucleus; this indeterminacy could be attenuated by making recourse to the knowledge of the chemical and physical properties of the elements, trying to reproduce the periodicity demonstrated in the Mendeleev table. The result at which Bohr arrived was that the number of electrons in the inner rings should increase with increasing atomic number; an erroneous result that in 1913, however, could not be recognized as such.

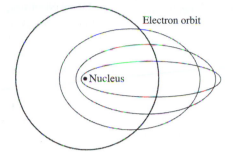

Figure 28. Some examples of electron orbits around the nucleus. In addition to circular orbits, elliptic orbits with various eccentricity are shown.

Bohr then turned his attention to molecules, obtaining correct results for the hydrogen molecule but surprisingly not so for the helium atom which also has two electrons.

Even if Bohr at the beginning of his work had referred to elliptical orbits, subsequently he focused exclusively on circular orbits. He also confined his discussion to the non-relativistic case by assuming that the electron velocity was small with respect to that of light. When in 1914 the American astronomer H D Curtis (1872–1942) found a small systematic disagreement between the theoretical values of the wavelengths of the hydrogen lines calculated by Bohr's theory and the experimental values, Bohr remade the calculations by introducing the relativistic variation of the electron mass. The correction moved in the right direction but was, however, too small to explain the observed deviations.

It was so that a Prussian physicist, Arnold Sommerfeld (1868–1951)—who had studied in Göttingen with the famous mathematician David Hilbert (1862–1943) and was very able in mathematics—in 1915 tried to improve the model extending calculations to the more general case in which electrons turn around the nucleus in elliptical and not just circular orbits, just as planets do around the Sun (figure 28). In this way the state of each electron in an atom was given by three numbers, called quantum numbers. These numbers were mutually linked by simple rules and characterized the energy of the electron in the orbit and the parameters of the orbit that allowed its form and orientation to be determined. The quantization criteria established that they could have only integer values. By a trick of the mathematics of the problem, even if the possible orbits of electrons were enormously increased in number, the possible energetic states remained the same. Many orbits with different parameters were travelled by electrons with the same energy, and this property, called degeneration, made it possible that the energy levels of the electron were still the ones that Bohr had calculated considering only circular orbits.

Sommerfeld was not discouraged and treated the problem relativistically, finding that the energy of the electron depended in this case also on

141

the shape of the orbit. Degeneracy was so eliminated and the result was in agreement with the spectroscopic observations already performed in 1891 by A A Michelson, who found that the lines of the hydrogen Balmer series are actually made by several lines very near to each other (fine structure). This fact was incompatible with Bohr's theory but initially had been ignored because the effect was very small.

Moreover a series of effects had to be explained. Besides the Zeeman effect, which has already been discussed, in 1913 Johannes Stark, in his laboratory at the Technische Hochschule of Aachen, discovered that an electric field can also split a spectral line of the Balmer series into several lines, and that this phenomenon is not peculiar to hydrogen.

> Johannes Stark (1874–1957) between 1906 and 1922 taught at Göttingen, Hannover, Aachen, Griefswald and Würzburg Universities. At this point his academic career arrested and notwithstanding his Nobel prize for physics, awarded in 1919 for his discovery, was rejected by six German universities. His unpopularity was due to his anti-Semitism that brought him to denounce the quantum theories and Einstein's relativity as the flawed product of 'Jewish science'. Having joined the Nazi party in 1930, although he was refused by the Prussian Academy of Sciences, in 1933 he succeeded in obtaining the presidency of the Imperial Institute of Physics and Technology. Here he tried to use his position of power to gain control of German physics but entered into conflict with the politicians and the administrators of the Ministry of Education of the Reich who, judging he was too destructive and unreliable, forced him to resign in 1937. The final humiliation came in 1947 when he was sentenced to four years of labour by a German court during the trials against Nazi Germany.

The effect of an electric field on spectroscopic lines was also independently discovered in Florence by Antonino Lo Surdo (1880–1949). Although the experimental set-up of Lo Surdo was much simpler than that used by Stark, he was only able to obtain qualitative results, having no direct means of making precise measurements in his laboratory. Stark strongly opposed calling the effect the Stark–Lo Surdo effect, refusing to give Lo Surdo any credit.

Immediately after this discovery, the German physicist Emil Warburg (1846–1931) and Bohr in 1914, using Bohr's model of the atom, presented an explanation of the effect which, however, agreed only qualitatively with the experimental results, that is it provided an understanding of why an electric field splits the energy level into several sub-levels but did not produce the correct value of the splitting.

In 1916, using the more refined model of elliptical orbits, P Debye (1884–1966), Nobel prize 1936 in chemistry, and Sommerfeld succeeded in giving an explanation of the normal Zeeman effect; however, the anomalous effect was still an enigma. The problem was still unresolved when Sommerfeld, around 1920, suggested using an empirical explanation by taking advantage of the experimental data. He derived the energy levels from the observed frequencies of the spectroscopic lines, finding then the quantum numbers that identified them and which allowed, via suitable selection rules, the transitions to be predicted.

Following this methodology, Sommerfeld introduced a new quantum number that he called the internal quantum number which was denoted (after Bohr's suggestion) by the letter j. A model was then developed, called the vectorial model, in which the number j represented the sum of the electron angular momentum vector and the angular momentum of the rest of the atom made by the nucleus and the inner electrons. The two moments summed according to complex quantum rules.

Meanwhile, A Landé (1888–1975) started to obtain the solution of the anomalous Zeeman effect but threw the situation into an incomprehensible state when he showed that in some cases the quantum numbers linked to the magnetic behaviour could have half-integer values. In all these still mysterious investigations, the idea that the electron orbit possessed a quantized spatial position appeared. The idea of spatial quantization was thus developing. In 1921, direct confirmation was given by Otto Stern (1888–1969) and Walther Gerlach (1889–1979).

Spatial quantization

Otto Stern, after earning his doctorate in 1912 at Breslau University, followed Einstein to Prague (1912) and Zurich (1912–14). In 1914 he became Privatdozent at Frankfurt-am-Main University. During the First World War he was a soldier in the German Army. He was then appointed professor in various German universities, and finally in Hamburg. Returning from the war, in Frankfurt he devoted himself to the development of the method of molecular rays begun by the Frenchman Louis Dunoyer (1880–1963). In this method, which requires a very high vacuum, a beam of free molecules or atoms is produced for use in experiments. The most important requirement is to perform the experiment in extremely clean conditions, similar to the ideal conditions suggested by theory. Stern and his collaborators investigated fundamental points concerning the kinetic theory of gases, verified spatial quantization, measured the magnetic moment of the proton, verified the de Broglie relation for waves made by helium atoms etc.

He left Germany in opposition to Hitler in 1933 and emigrated to the United States where he went to Carnegie Institute of Technology, Pittsburgh. He was awarded the Nobel prize for physics in 1943 for his research on molecular beams.

The first work Stern did with molecular beams was to produce direct evidence for Maxwell's velocity distribution law and to measure the mean velocity of molecules.

Max Born (1882–1970), who went to the University of Frankfurt in 1919 as professor of theoretical physics, recalled he was so fascinated by the idea of this measurement that he put all the means of his laboratory, workshop and mechanics at Stern's disposal. Stern was not very good with his hands in the laboratory, but he knew quite well how to direct a technician who did everything. Later, in 1920, Walther Gerlach, an excellent experimentalist, came to Frankfurt, and Born was very happy at this news. 'Thank God, now we have someone who knows how to do experiments. Go on, man, give us some help!' he said. Walther Gerlach received his doctorate in Physics at the University of Tübingen in 1912. While serving in the military during the First World War, he worked with Wilhelm Wien on the development of wireless telegraphy. After a brief interlude in industry, he came in Frankfurt.

In Frankfurt, Gerlach, who also had worked on atomic beams when he was with Friedrich Paschen in Tubingen, designed an experiment to study the deflection of a beam of bismuth atoms in a non-homogeneous magnetic field to determine its magnetic properties.

One day, Stern came to him and said, 'With the magnetic experiment one can do something else. Do you know what directional [space] quantization is?', and Gerlach replied 'No, I don't know anything about it'. Many physicists at that time did not believe that spatial quantization really existed, and thought it was only a way to perform calculations. Gerlach later remembered that Peter Debye had remarked to him: 'But surely you don't believe that the [spatial] orientation of atoms is something physically real; it is [only] a prescription for calculation', and even Born was of the same opinion. Stern on the contrary believed it was a representation of reality and so, after having explained the effect to Gerlach told him: 'It is worth trying' and proposed 'Shall we try? Let's prove it'.

The original proposal by Stern was detailed in a paper by the title 'A method of testing experimentally the directional quantization in a magnetic field':

> 'in the quantum theory of magnetism and the Zeeman effect one
> assumes that the angular moment vector of an atom may form with
> the direction of the magnetic field strength H only discrete and well
> determined angles, such that the angular moment in the direction of H
> be an integer multiple of $h/2\pi$'.

To understand this statement it is necessary to remember that already André-Marie Ampère (1775–1836)—the French physicist who gave the mathematical basis to electromagnetism, finding the connection between electricity and magnetism, and who for his talent was named by Napoleon (1808) general inspector of the new French university system—had demonstrated that an electric current in a circuit generates a small magnetic moment as if the circuit was an elementary magnet. In atoms, the orbits travelled by

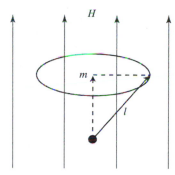

Figure 29. Precession of angular momentum *l* in a magnetic field *H*. The projection *m* of *l* along the *H* direction is shown.

electrons revolving around the nucleus may be assimilated to small coils through which currents pass, and Bohr's quantum mechanics, improved by Sommerfeld, allows the calculation of the magnetic moments associated with each orbit. These sum via complex rules found by Sommerfeld so that many atoms possess a magnetic moment and therefore behave as the needle of a compass which may be oriented by an external field. If an atom is subjected to an external magnetic field, its magnetic moment, referred to by the letter *l*, according to the rules of electromagnetism and classical mechanics tries to align parallel to the external field. To do this it describes a cone that has an axis along the direction of the field (precession motion). Its projection along the direction of the field (let us indicate it with *m*) is what we call the component of the moment along the field (figure 29). Now it happens that while according to classical mechanics the *m* component along the field may have any possible value between $+l$ and $-l$ (that is any angle between *l* and the field is allowed) according to quantum theory only discrete values of *m* are possible, corresponding to $m = l, (l-1)$, $(l-2), \ldots, -l$ (that is only some angles of *l* with respect to the field are allowed) (figure 30). According to quantum mechanics the magnitude of the vector *l* is $\sqrt{l(l+1)}$ which is greater than the larger value of *m*. It is therefore clear that l can never point exactly in the field direction, a circumstance that is intimately linked to the uncertainty principle of Heisenberg. To clarify this concept in a simple way for the case in which the magnetic moment of the atom expressed in suitable units is $l = 1/2$, one may think that the magnetic field is oriented from bottom to top as shown in figure 31 and that the atom is a man who is holding an arrow (which in our figure is its magnetic moment). While according to classical laws the man may orient the arrow in any direction, according to quantum mechanics only the two positions shown in the figure are allowed (we will call them the parallel and the anti-parallel orientations with respect to the field) in which the projection of the arrow in the field

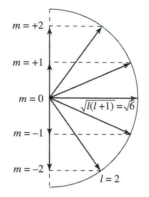

Figure 30. Possible orientations of the angular momentum *l* along the direction of the external magnetic field *H* are shown. In the figure, $l = 2$ (in suitable units) and the corresponding m values are ± 2, ± 1 and 0. There are therefore five possible orientations of *l* that are shown by the arrows.

direction is $+1/2$ (figure 31a) or $-1/2$ (figure 31b). The arrow has therefore a length that is $\sqrt{3/2}$.

To verify that atoms may orient themselves only in a discrete manner, Otto Stern designed an experiment based upon the deflection of a molecular beam in a non-homogeneous magnetic field. We said the atom with its magnetic moment is like a small magnet. If we make it travel in a region where a homogeneous magnetic field exists, it will be acted upon by a magnetic force on its north pole with the same strength as on the south pole, but in exactly the opposite direction so its magnetic moment is oriented in the direction of the external field (the precession motion we discussed previously), but no force deviates its motion.

If the field is not homogeneous this is not the case. Here, the forces acting on the north and south poles are not quite the same, so that, besides the orienting action, a resultant force also acts on the magnet as a whole which deviates

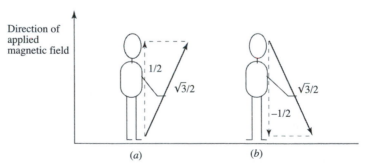

Figure 31. An atom with momentum $l = \frac{1}{2}$ (in suitable units) may orient its momentum only in the two possible ways shown in the figure by the pointer in the man's hand.

146

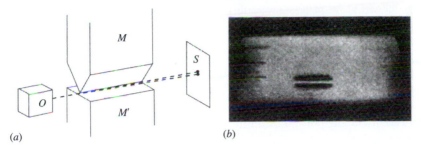

(a)   (b)

Figure 32. (a) Layout of the Stern and Gerlach experiment. A molecular beam exits from an oven O and travels through the poles of a magnet MM' (which is shaped like a knife) ending its trajectory on the screen S. (b) Magnetic splitting of a lithium atom beam.

the atom from its original trajectory. The amount of deflection is determined by the degree of the inhomogeneity of the field. In fact to produce an appreciable deflection the inhomogeneity of the field must be so marked that the field changes decidedly even within the small length scale of the elementary magnet (which in our case is of the order of magnitude of atomic linear dimensions, that is about one hundred millionth of a centimetre). Stern succeeded in producing a sufficient degree of inhomogeneity by suitable construction of the pole-pieces of the magnet, one piece being shaped as a knife-edge, while the other piece, which he set opposite it, had a flat face (figure 32a). In this configuration the magnetic field is much stronger near the knife-edge than at the other pole-piece. A fine beam of atoms, obtained by vaporizing in an oven the substance whose atoms were under study, was projected through two circular holes to collimate it well and passed through the pole-pieces. Each individual atom was deflected in the non-homogeneous field, according to the magnitude and direction of its magnetic moment. The traces of the individual atoms were made visible on a suitable screen. The experiment was very difficult because the whole system must work in a high vacuum necessary not to disturb the motion of the atomic beam and yet at that time the vacuum pumps were very delicate and often broke. It took almost two years from 1921 to 1922 to finish the entire experiment.

According to classical theory, the atoms should deviate in all possible directions because their moments may have any orientation with respect to the field. Therefore on the screen a large spot should be observed, produced by the incident beam. Quantum theory, on the contrary, predicts that directions are quantized and only a discrete number of orientations are possible. The mark on the screen is therefore split up into a finite number of discrete beams. The experiment made with silver atoms showed nothing at first. Stern described it in an interview:

> 'After venting to release the vacuum, Gerlach removed the detector flange. But he could see no trace of the silver atom beam and handed the flange to me. With Gerlach looking over my shoulder as I peered

147

closely at the plate, we were surprised to see gradually emerging the
trace of the beam.... Finally, we realized what [had happened]. I was
the equivalent of an assistant professor. My salary was too low to
afford good cigars, so I smoked bad cigars. These had a lot of sulfur in
them, so my breath on the plate turned the silver into silver sulfide,
which is jet black, so easily visible. It was like developing a
photographic film.'

And so finally the plate showed that the beam had split up into two separate
sub-beams!

But the result was not very clear. The experiment continued despite big
financial difficulties due to the situation in Germany and Born started a series
of public lectures on relativity, charging an entrance fee which supported the
experiment.

Later on, Stern accepted a professorship in Rostock, and Gerlach, left
alone, repeated the experiment by substituting the circular holes that created
the beam with a rectangular opening. In this way the number of atoms that
passed was much greater and the image obtained was much clearer. In figure
32b the result of the experiment using lithium atoms is shown. The result is
very clear. Instead of one broadened central spot, two lateral well separated
spots are obtained. This not only demonstrates that atoms have a magnetic
moment which corresponds to an angular moment whose projection is $+1/2$
or $-1/2$, but also allows its value in absolute units to be measured. Of course
the interpretation that Stern gave at the time was not fully correct because the
spin of the electron (we will speak of this in a moment), which generates a
magnetic moment that must be summed with that produced by the electron
in its orbit using quantum mechanical rules, was not taken into account.
However, the basic principle of the quantization of the magnetization direction is true in any case. Many problems that arose in this experiment were
solved when the spin of the electron was discovered. For these experiments
Stern was awarded the physics Nobel prize in 1943. Later Stern came to
the USA and in 1945 retired and settled in Berkeley, California.

Gerlach in 1925 returned to Tübingen as professor and then came to
Munich. During the war in 1944, he became Head of the German Nuclear
Research Programme and at the end of the war was among the ten leading
scientists detained at Farm Hall by Allied forces. Later he contributed
much to the rebuilding of German science and campaigned to ban nuclear
weapons.

The exclusion principle

Notwithstanding its apparent successes, in 1924 the 'old' quantum theory,
which a few years earlier seemed to have produced methods and principles
able to help at least in the task of representing the fundamentals of atomic
phenomenology, appeared in difficulty. At this moment, Wolfang Pauli

(1900–1958), reflecting on the embarrassments of the theory, found the starting point for the introduction of a new and mysterious principle.

Pauli was born in Vienna, the son of a physician who became professor of biochemistry in the local University. He joined the University in Munich under the guidance of Sommerfeld and after taking his degree for a year was an assistant of Born in Göttingen in 1921–22. Having met Bohr, on his invitation, he went to Copenhagen from October 1922 to September 1923 and then to Hamburg, where he remained until 1928 when he accepted a chair at the Polytechnic in Zurich. Except for the war years, which he passed in the United States at the Institute of Advanced Study in Princeton, he remained in Zurich until his death in 1958.

Pauli was a large, colourful and jovial person. When he appeared, his resounding and somehow sardonic laugh animated any meeting, however tedious it was at the start. He always arrived with new ideas. He began his scientific career by writing at 21 a book on Einstein's relativity theory that still today is one of the best on the subject. He made very important contributions to quantum theory. Like all theoreticians he had no gift for any contact with experimental instrumentation. It is said he also possessed a mysterious power. Once, in Professor Frank's laboratory in Göttingen, without any apparent cause, a complex device for the study of atomic phenomena broke. Frank, amused, wrote of the fact to Pauli in Zurich and after some time received his reply in an envelope with a Danish stamp. Pauli wrote that he had gone by train to pay a visit to Bohr and at the very moment of the accident in Frank's laboratory his train stopped for a few minutes in Göttingen station.

Pauli, as many people said, was the conscience of physics. He wanted people to understand things thoroughly and expressed them correctly. He never got tired of answering questions and explaining problems to anybody who came asking, but he had no difficulty in expressing his dislike when he thought somebody was saying something wrong. He was not a good lecturer, because he did not have the ability to judge how much the audience could take in. Once a student dared to interrupt him and said: 'You told us that conclusion is trivial, but I am unable to understand it'. Then Pauli did what he did frequently when he had to think things over during a lecture: he left the room. After a few minutes he came back and said: 'It is trivial' and resumed the lecture.

Once his assistant, the Vienna nuclear physicist Victor Weisskopf (1908–2002) published a paper in which there was an error and Pauli consoled him saying: 'Don't take it too seriously, many people published wrong papers; I never did!'. Still Weisskopf one day showed Pauli a newly published paper on a subject of his interest. He said: 'Yes, I thought of that too, but I am glad he worked it out, so that I don't need to do it myself'.

At the time of our story, one day a colleague encountered Pauli wandering aimlessly about around the Copenhagen streets. It was at the time he paid a visit to Bohr and became interested in the Zeeman effect. The colleague said to him, 'You look very unhappy', and Pauli replied:

'How can you be happy while you think about the anomalous Zeeman effect?' However, this thought eventually bore important fruit as in 1924, Pauli, giving up with a mechanical vision of the atom and focusing attention only on the quantum numbers that represented the electron states, pronounced, on the basis of a long study of the Zeeman effect, that each electron is characterized by a set of quantum numbers and that in an atom no more than two electrons with the same quantum number were allowed to exist. In simpler words this means that no more than two electrons can travel in each orbit around the atom. This principle that more than two electrons are excluded from a given orbit was termed by Pauli the 'exclusion principle'. Immediately, this rule allowed electrons to be assigned to various energy levels and the Mendeleev periodic table to be built up. The principle was later demonstrated using quantum mechanics and Pauli was awarded the Nobel prize for physics in 1945.

The spin of the electron

The final modification to the old quantum theory which further explained experimental observations was provided in November 1925 by G E Uhlenbeck (1900–1988) and S A Goudsmit (1902–1978) who discovered that the electron revolves around its centre like the earth, and is similar to a small elementary magnet. The quantity that characterizes the rotation of the electron is called its spin, while that characterizing its magnetic properties is its magnetic moment. Uhlenbeck and Goudsmit found that the electron spin in suitable units has the value $h/4\pi$. At the time they were just over 20 years old, and their discovery was the consequence of an accurate study of atomic spectra.

> George Eugene Uhlenbeck was born in Batavia (now Jakarta), the son of a Dutch military man, who remained for some time in the Dutch Indies (now Indonesia). Therefore the young George attended elementary school in Sumatra. In 1907 the family moved permanently to Holland and settled in the Hague. In 1919 Uhlenbeck enrolled in the University of Leiden to study physics and mathematics with Paul Ehrenfest, the Dutch physicist Heike Kamerling Onnes (1853–1926)—an expert on low temperatures, the first person able to liquefy helium and the discoverer of superconductivity, being awarded the Nobel prize in 1913—J P Kuenen and H A Lorentz. Between 1922 and 1925 he visited Rome, where he was the private tutor of the younger son of the Dutch ambassador. During his stay he studied Italian, became a friend of E Fermi (1901–1954) and was deeply involved in history. When he left Rome for good to return to Holland he was seriously considering giving up physics to become a historian. He discussed the matter with his uncle, the distinguished linguist Christianus Cornelius Uhlenbeck, an expert on American Indian languages and a professor at Leiden. His uncle was sympathetic to the idea but suggested that it might be best to first obtain a PhD in physics, because he had already progressed quite far. Ehrenfest also

agreed to this idea but proposed that first he should find out what was currently happening in physics. Therefore he encouraged Uhlenbeck to work with him for some time and to learn from Goudsmit what was going on in *Spektralzoologie,* as Pauli used to refer to the study of spectra.

Samuel Abraham Goudsmit was born in the Hague, the son of a well-to-do merchant and discovered his interest for physics at the age of 11 after reading a physics text. He was particularly struck by a passage explaining how spectroscopy shows that stars are composed of the same elements as the Earth. At University he studied with Ehrenfest demonstrating an intuitive rather than analytical way of thinking. Uhlenbeck later said 'Sam was never a conspicuously reflective man, but he had an amazing talent for taking random data and giving them direction. He's a wizard at cryptograms'. I I Rabi added 'He thinks like a detective. He is a detective' and in fact Goudsmit once took an eight-month course in detective work.

In 1920, Ehrenfest recommended Goudsmit to visit Paschen in Tubingen who described his research on spectroscopy. The following year Goudsmit returned to Tubingen during the summer and Paschen introduced him to spectroscopic techniques. Soon he became very capable of handling quantum numbers in order to explain the observed spectra. Early in 1925 he published a paper in which he showed that it was possible to simplify the application of the Pauli exclusion principle using the Landé half-integer quantum numbers and that one of these numbers always had the value $+1/2$ or $-1/2$. At this time, Ehrenfest asked Uhlenbeck and Goudsmit to work together; Goudsmit to explain to Uhlenbeck the magic calculations with quantum numbers and Uhlenbeck to teach Goudsmit some physics in order to show him that it did not consist only of the manipulation of quantum numbers.

In August 1925 the two men, who became linked by a friendship that lasted throughout their lives, began their regular meetings in the Hague and one morning towards the end of the summer Goudsmit spoke to Uhlenbeck about the Pauli principle using the half-integer quantum numbers of Landé. Immediately Uhlenbeck observed that all electrons behave as if, besides turning around the atomic nucleus, they could also rotate around themselves (spin). In September the theory was completed and the two researchers showed that this concept also explained the normal and the anomalous Zeeman effect.

The Hungarian–American physicist R de L Kronig (1904–1995), who had travelled to Italy and was also a friend of Fermi, had already formed the same idea concerning spin. Kronig had the misfortune to ask Pauli for his opinion, and Pauli persuaded him that his hypothesis was deprived of any foundation and ridiculed the idea saying of it 'that it is indeed very clever but of course has nothing to do with reality'. So Kronig gave up on it. When Uhlenbeck and Goudsmit learnt of Pauli's criticisms, which seemed valid, they considered withdrawing the paper already sent for publication but Ehrenfest told them that since they were so young they could allow themselves to make a mistake. One objection, for example, was that if one uses the formula given by H A Lorentz for the size of the electron, then so fast a rotation is needed to have a self-rotating angular momentum of 1/2 that the electron's surface reaches a speed ten times higher than that of light. The work was not withdrawn and was later published and Pauli's criticism remained groundless.

> After the great discovery of spin, in 1927 Uhlenbeck emigrated to America to the University of Michigan. In the mid-1930s he came back to Holland where he was appointed the successor of Kramers in the University of Utrecht, only to return to Michigan University in 1939. In 1960 he went to the Rockfeller University in New York. He was a foreign member of the Italian Academy of Lincei.
>
> Goudsmit too emigrated to the University of Michigan in 1932. During the Second World War he worked on radar and later in 1944 he was the head of a very secret mission code-named Alsos. The mission would follow the advance of the allied forces in Europe and in some cases precede them to discover the progress made by the Germans in the building of the atomic bomb. He found that German scientists did not achieve a great deal and it was clear that Hitler never had the weapon before the end of the war. Goudsmit wrote the story of the mission in the book *Alsos* (1947).

As a final comment we may observe that the full development of quantum mechanics that occurred in the years following provided an adequate treatment of the behaviour of atoms and molecules. For us, however, what has been described up to now in this book is sufficient to understand the main facts. We may imagine atoms and molecules as complex systems that may exist in several energy states. The simpler system, the atom, owes its energy states to the arrangement of its electrons. The energy difference between electronic orbits corresponds to photons emitted in the visible and ultraviolet regions. However, the energy corresponding to a given orbit can be perturbed by different causes due to the interaction of the magnetic moment of the electron (due to its spin) and the magnetic moments produced by its motion around the nucleus or the magnetic moment of the nucleus or by the action of external magnetic fields (Zeeman effect) or electric fields (Stark–Lo Surdo effect). The result of these interactions is that the energy level of the unperturbed orbit splits into several sub-levels that differ slightly in energy. The transitions that may occur among these sub-levels correspond to the so-called fine structure or hyperfine structure and the corresponding wavelengths are in the infrared or microwave region.

Molecules are more complex systems composed of atoms. Besides the electronic levels, they have other energy levels due to the fact they have rotational motion and that the constituent atoms vibrate around their equilibrium positions. According to quantum mechanics, the energies corresponding to the rotational and vibrational motions are also quantized. Thus, for any electronic configuration a series of energy levels that we may call roto-vibrational levels exist. The energies corresponding to the jumps between roto-vibrational levels for the same electronic configuration are very small. The corresponding wavelengths are in the infrared or microwave regions.

Therefore, in the spectrum of any substance, all the lines in the visible and ultraviolet region arise due to electronic transitions, while the lines in the infrared and microwave region mostly originate from transitions between roto-vibrational levels, or between sublevels of fine and hyperfine structures, or are produced by Zeeman or Stark–Lo Surdo effects. This rule is not entirely stringent because the energies corresponding to transitions between very highly excited electronic levels (that is electrons very far from the nucleus) may also be very small and correspond to wavelengths in the infrared or microwave range. In what follows, however, we will not consider these cases.

CHAPTER 9

MAGNETIC RESONANCE

We have seen that the motion of any particle, atom or molecule possessing a magnetic moment is influenced by an external magnetic field. For the sake of a simple representation we may consider the magnetic moment of our particle as an arrow that points in some direction. The external magnetic field exerts on the magnetic moment of the particle, that is on our arrow, a pair of forces that would rotate and align it in the direction of the field. However, if the particle rotates around itself, as the Earth does around its axis or as the electron does due to its spin, the presence of the rotation profoundly changes the effect of the forces and thus the magnetic moment of the particle begins to rotate around the external field with an angular velocity (proportional to the magnetic field) referred to as the Larmor frequency (named after an Irish physicist who discovered this phenomenon). This type of motion is called Larmor precession. The motion is similar to that of a spinning top which turns around itself with its axis inclined with respect to the vertical direction: the rotation axis of the spinning top turns slowly around the vertical direction (figure 33) during its precession motion.

In the case of an atom or molecule, its magnetic moment cannot have *any* inclination with respect to the external field because due to quantization only certain inclinations are possible (see figure 30). The magnetic moment of the particle may only perform rotations around the external field, at their Larmor frequency, which correspond to the values of the allowed inclinations. To each of these motions and therefore to any of these inclinations (angles) corresponds a well defined energy and therefore in order to change from one inclination to another it is necessary to increase or decrease the particle energy by the energy difference between the two inclinations, or as we will say between the two energy levels.

If the total angular moment of the particle is $\frac{1}{2}$ in suitable units the particle may only be aligned to the field, nearly parallel or nearly anti-parallel to it; if its angular moment is different from $\frac{1}{2}$, the number of possible directions increases, as shown in figure 34.

By using this property, Stern and Gerlach gave the first experimental proof of space quantization, being able to measure the angular moment of certain atoms and, by improving the technique, in a series of experiments

154

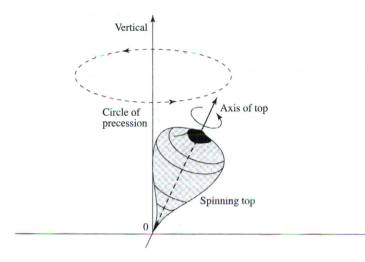

Figure 33. A spinning top rotating around its axis, inclined with respect to the vertical, performs a precession motion around the vertical direction.

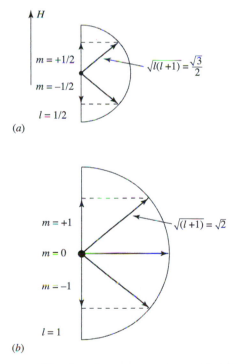

Figure 34. In (a) the two possible orientations of an angular momentum $l = \frac{1}{2}$ (in suitable units) with respect to the direction of an external magnetic field are shown. In (b) the three possible orientations of a momentum $l = 1$.

155

performed with co-workers between 1933 and 1937, Stern succeeded in measuring the magnetic moment of the proton and of the deuteron (the nucleus of the atom of heavy hydrogen, called deuterium, made by one proton and one neutron).

The resonance method with molecular beams

If suitable radiation, with a frequency that corresponds exactly to the energy difference between two energy levels (that is, at resonance, as one usually says) is incident upon a particle so as to make it to jump from one magnetic level to another one with a higher energy, the radiation is absorbed. Around 1932, the Italian physicist Ettore Majorana (1906–1938), who disappeared in mysterious circumstances at sea between Palermo and Naples, and Isidor I Rabi (1898–1988) discussed theoretically the absorption that occurs at magnetic resonance. G Breit (1899–1981) and Rabi had already theoretically predicted the possibility of also applying these techniques to the measurement of magnetic spin (1931) and of magnetic moments.

> Rabi was born in Poland, but his parents emigrated to the United States when he was still a boy and he grew up in the Jewish community of New York where his father owned a drugstore. In 1927 he earned a PhD from Columbia University to which, after two years spent in Europe, he returned for the rest of his life until he retired in 1967. In 1927 in Germany, Rabi worked with Otto Stern and his attention was attracted by the experiment Stern did with Walther Gerlach; therefore, when he came back to Columbia he continued to work on atomic and molecular beams, inventing a method of magnetic resonance we will describe shortly. Through using this method, after the Second World War he was able to measure the electron's magnetic moment with an exceptional accuracy, offering a very good test to check the validity of quantum electrodynamics. The method had huge applications in atomic clocks, in nuclear magnetic resonance and therefore in masers and lasers. During the Second World War he worked on the development of microwave radar.

In a famous paper written in 1937, Rabi described the fundamental theory for magnetic resonance experiments. At this time in his laboratory the magnetic moments of many nuclei were under measurement with the method of the inhomogeneous magnetic field used by Stern. The measurements started in 1934 and continued until 1938. Rabi, however, wanted to improve them and therefore studied the effect of the precession motion of the spin around a magnetic field; but didn't pay attention to the resonance phenomenon that can be produced if radiation that has exactly the frequency corresponding to the energy difference between one level and the other is sent onto the spin. In September 1937, C J Gorter (1907–1980), then at Gröningen University in Holland, paid a visit to Rabi and described to him his unsuccessful attempts to observe nuclear magnetic resonance effects in solids. In

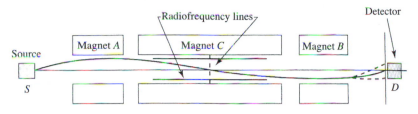

Figure 35. Layout of a molecular-beam magnetic resonance experiment. The beam from a source crosses two regions, *A* and *B*, with inhomogeneous magnetic fields, which deflect the beam in opposite directions. If the molecule does not change its spin state as it comes through region *C*, it has no net deflection. An oscillating field is introduced into region *C*. When the frequency of the oscillator is the same as the Larmor precession frequency it changes the spin orientation of the molecule and the beam intensity at the detector drops sharply. (From Ramsey N F October 1993 *Phys. Today* 41).

the course of his discussion with Gorter, Rabi began to appreciate the resonant nature of the phenomenon and immediately with his collaborators modified his instrumentation. So in 1939 the method underwent a notable improvement that allowed the reorientation of the atomic, molecular or nuclear moments with respect to a constant magnetic field, superposing an oscillating magnetic field. When the frequency of the oscillatory field is equal to the energy difference between two levels in the magnetic field divided by Planck's constant, a reorientation may occur that in this case is resonant and may involve an absorption from a lower to a higher level or, in competition a stimulated emission process, from a higher to a lower level. To detect these reorientations, Rabi and his collaborators, J M B Kellog, N F Ramsey and J R Zacharias used an ingenious system consisting of two regions in which an inhomogeneous magnetic field acted on the beam (figure 35). In the first region (*A*) the field deflected the molecular beam in one direction, while in the second region (*B*) the inhomogeneous field was applied in the reverse direction, deflecting the beam in the opposite direction so as to refocus it on a detector. Because the focusing and defocusing effects depend in the same way on the velocity of the molecules, in the end all molecules, regardless of their velocity were refocused on the detector. A strong magnetic field was applied (*C*) in the centre of the structure that produced a Larmor precession of the magnetic moments of the molecules and in the same region a weaker alternating magnetic field was superposed. If the frequency of this alternating field was equal to the Larmor frequency, it was able to reorient the magnetic moment of the molecule and therefore the particle in the second region was deflected in a different way and was no longer focused. So if the intensity of the beam is observed, by making the intensity of the strong magnetic field constant and slowly changing the frequency of the weak alternate field, a curve such as the one shown in figure 36 is obtained, which shows a minimum in the signal from the detector

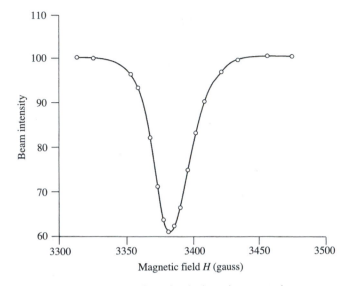

Figure 36. A typical resonance curve of a molecular beam in a magnetic moment measurement. (From Rabi I S *et al* 1939 *Phys. Rev.* **55** 526).

when the frequency of the oscillation is equal to the frequency corresponding to the jump between the two levels.

For these experiments Rabi was awarded the Nobel prize in 1944. He was highly reputed in the scientific world and many people asked his advice on how to direct their own research. In 1937 he discouraged Ramsey, then his PhD student at Columbia University, to continue research on molecular beams because there was little future in the field. Ramsey boldly ignored the advice of his master and as we will see provided important contributions in that area, some of which found application in the building of atomic clocks. Yet Rabi himself, seven years later, earned the Nobel prize just for the development of the beam-resonance method.

> Norman F Ramsey was born in Washington, DC, in 1915. His father was an officer in the Army Ordnance Corps and his mother, daughter of German immigrants, had been a mathematical instructor at the University of Kansas. He entered Columbia College, New York, in 1931 and graduated in 1935, then to enrol at Cambridge University, England, as a physics undergraduate. In the summer of 1937, after two years at Cambridge, he returned to Columbia to work with Rabi. Rabi had just invented his method and Ramsey was the only student to work with him and his colleagues on one of the first experiments.

In 1949, then at Harvard, while studying how to improve Rabi's method to perform measurements with greater accuracy, Ramsey invented the method of separated oscillatory fields in which the single oscillating magnetic field in the centre of the Rabi device is replaced by two oscillating fields at the entrance and exit, respectively, of the space in which the nuclear magnetic moments are to be investigated. The method offered a number of advantages and found many applications.

A very similar method to the one introduced by Rabi was used by LW Alvarez (1911–1988) and F Bloch in 1940 for measuring the magnetic moment of the neutron, which they determined within an accuracy of 1%. Because their papers describing the experiment appeared three years after the first magnetic resonance papers from Rabi, their technique is usually presumed to be an adaptation of Rabi's method. However, Bloch had the idea independently to use magnetic resonance by employing an oscillating magnetic field. Alvarez was awarded the Nobel prize in 1968 for the development and use of a device for the detection of elementary particles in nuclear physics (the bubble chamber). At the end of the 1970s he was the author of a bold hypothesis that linked the extinction of prehistoric animals, 65 million years ago, to the cataclysm produced by an impact of a gigantic meteorite with the Earth. A huge cavity of 180 km diameter was later discovered in the region of Chicxulub in the Yucatan peninsula (Mexico) that is thought to have occurred due to the impact with an extraterrestrial object in approximately the same period. Much evidence for Alvarez's hypothesis has been obtained in the ensuing years.

Magnetic relaxation phenomena in solids

If we now consider a solid material, the different magnetic moments of nuclei or of electrons, in the presence of an external magnetic field, may sum up to give a total magnetic moment (this happens in certain substances called paramagnetic materials) that produces a gross magnetization of the material.

The problem of how fast the mean magnetic moment of a paramagnetic substance responds to a sudden change in the magnetic field in which the substance is placed, had already been treated in the 1920s by W Lenz (1888–1957), P E Ehrenfest, G Breit and Kamerling Onnes, and in the 1930s great attention was given to the way in which a magnetic system reaches thermal equilibrium.

The Swedish physicist Ivar Waller (1898–1991), in a now famous paper which appeared in 1932, had already distinguished the two main mechanisms through which the system reaches equilibrium after being disturbed (relaxation phenomena): the interaction of the spins of electrons or nuclei with each other, that we will call spin–spin interaction, and the mechanism of interaction among the spins and the atoms that form the solid (the lattice as it is called) that we will call spin–lattice interaction. If the interaction of spins with each other and with the lattice did not exist, the application of a static field would produce only the precession motion of individual spins, independent from one another, without producing a collective motion.

Conversely, the effect of the interactions of spins with each other and with the lattice and the presence of the external magnetic field produce energy exchanges between spins and amongst spins and the lattice and, because atoms with different orientation of their spins may assume different energy levels, a distribution of levels results. We may therefore apply Boltzmann's formula, used in statistical mechanics, according to which the states corresponding to lower energy are more probable. After a relaxation time (which is longer the smaller the interactions are), a thermodynamic equilibrium is reached in which the probability of finding a nucleus or an atom with its spin parallel to the field (lower energy state) is greater than the probability of finding it with anti-parallel spin.

The important concept of spin–lattice relaxation was later taken up again in 1937 by H B G Casimir (1909–2000) and F K Dupré. They noted that concerning the interaction among spins and the lattice, one may consider that electrons, being light and fast, interact strongly and rapidly with each other, reaching a spin equilibrium corresponding to some temperature, in a very short time of about a tenth of millisecond (spin–spin relaxation time). Then the spin system starts to interact with the more massive lattice that in general is at a different temperature from the electron spin system, and reaches equilibrium in a much longer time, typically of the order of a millisecond or even longer (spin–lattice relaxation time). The diversity of the times in which the spin–spin and spin–lattice interactions take place allows the magnetic crystal to be divided into two sub-systems, one formed by the spins, and the other by the lattice, each one with its own temperature.

Excitation of magnetic levels was at this time carried out in order to study atomic levels and nuclear spins. The Dutch physicist C J Gorter had considered in the late 1930s and early 1940s the possibility that nuclear magnetic moment precession in an external field could give rise to macroscopic effects. In 1936 he attempted to detect nuclear resonance in solids by looking for an increase in temperature, and showed remarkable insight by attributing the negative results of his experiment to a long spin–lattice relaxation time (in his case larger than a hundredth of second). He discussed these experiments with Rabi in 1937.

Magnetic resonance

The first successful experiments to detect magnetic resonance in matter by electromagnetic effects were carried out independently by F Bloch at Stanford, E M Purcell at Harvard and E Zavoisky in the former USSR. In these experiments the reorientation of the magnetic moments of nuclei or of electrons in a solid via interaction with an electromagnetic wave was obtained. This technique is called magnetic resonance.

To observe the effect, two magnetic fields are needed: one static field to create the energy states and one oscillating field that induces the transitions between the states. In this way absorption or emission of radiation occurs that produces changes in the distribution of the energy levels. The phenomenon is in some sense analogous to transitions of electrons between their orbits that yield spectra in the visible, infrared and ultraviolet regions, observed in atoms, but is much more complex. In the case of visible, infrared and ultraviolet radiation, the levels among which the transition occurs exist already because they are the energy levels of the electron in the atom. In the magnetic case, first one needs to create the energy levels by means of some external field, and then the transitions can be studied.

Bloch and Purcell were awarded the Nobel prize in 1952 for their research. They introduced magnetic resonance through two different paths which were, however, substantially similar. Zavoisky was the first to observe transitions between fine-structure levels of the fundamental state in paramagnetic salts (paramagnetic electronic resonance).

Felix Bloch was born in Zurich, Switzerland, on 3 October 1905. He entered the Federal Institute of Technology (Eidgenossische Technische Hoschule where Einstein had also studied) in Zurich in 1924. After one year's study of engineering he decided instead to study physics and changed to the Division of Mathematics and Physics at the same institution. During the following years he studied under Professors P J W Debye (1884–1966), P Scherrer (1890–1969), H Weyl (1885–1955) and Schrödinger. He was interested initially in theoretical physics. After Schrödinger left Zurich in the autumn of 1927, he continued his studies with Heisenberg at the University of Leipzig, where he received the degree of Doctor of Philosophy in the summer of 1928 with a dissertation dealing with the quantum mechanics of electrons in crystals and the theory of metallic conduction with which he established the theoretical basis for the modern treatment of electrons in solids via band theory. Subsequently, he worked with Pauli, A H Kramers, Heisenberg, Bohr and Fermi.

After Hitler's ascent to power, Bloch left Germany in 1933. A year later he accepted a position at Stanford University, California. There he started experimental research which lasted until his retirement in 1971. In 1936 he published a paper in which he showed that the magnetic moment of free neutrons could be measured by the observation of their scattering in iron. During the war he also was engaged in the early stages of work on atomic energy at Stanford University and Los Alamos and later in counter-measures against radar at Harvard University. Through this latter work he became acquainted with modern developments in electronics which, towards the end of the war, suggested to him, in conjunction with his earlier work on the magnetic moment of the neutron, a new approach to the investigation of nuclear moments in solids. In 1945, immediately after his return to Stanford, he began the study of nuclear induction, as he was later to call it.

He also held important scientific positions. In 1954 he was the first Director General of CERN in Geneva, the large European organization for high-energy research. He died in Zurich on 10 September 1983.

Edward Mills Purcell was born in Taylorville, Illinois on 30 August 1912. In 1929 he entered Purdue University in Indiana where he graduated in electrical engineering in 1933. His interest had already turned to physics and K Lark-Horovitz, the great professor to whom solid-state physics in the USA is so indebted, allowed him to take part in experimental research in electron diffraction. After one year spent in Germany at the Technische Hochschule, Karlsruhe, where he studied under Professor W Weizel, he entered Harvard University, and received a PhD in 1938 working with K T Bainbridge. After serving two years as instructor in physics at Harvard, he joined the Radiation Laboratory at MIT, which was established in 1940 for military research and the development of microwave radar. He became head of the Fundamental Development Group in the Radiation Laboratory which was concerned with the exploration of new frequency bands and the development of new microwave techniques. The discovery of nuclear resonance absorption, as he called it, was made just after the end of the war and at about that time Purcell returned to Harvard as Associate Professor of physics. He became Professor of physics in 1949 and died in 1997.

Eugenii Konstanovich Zavoisky was born in Kazan. The son of a medical doctor, he studied and then worked at Kazan University. He was interested almost from his student days in the use of radio-frequency electromagnetic fields for the study of the structure and properties of matter. Beginning in 1933 he performed exploratory experiments on the resonant absorption of radio-frequency fields by liquids and gases. In 1941 he became the first to use the modulation of a constant magnetic field by an audio-frequency field in such experiments. In 1944 he discovered electron paramagnetic resonance, which became the subject of his doctoral dissertation. During the years 1945–47 he performed a series of important experiments, recording paramagnetic dispersion curves in the resonance range and obtaining electron paramagnetic resonance in manganese. Later he became associated with the Kurchatov Institute of Atomic Energy in Moscow where he worked for more than 20 years.

He made contributions to various fields of nuclear physics, developing among other things the scintillator-track chamber in 1952. In the area of plasma physics he discovered magneto-acoustic resonance in 1958. He was awarded the Lenin and State prize. He died in 1976. His studies became known in the West only after the Second World War.

The announcements of the first experiments on magnetic resonance were made independently by Bloch and Purcell within a month of each other. In the January 1946 issue of the prestigious American journal *Physical Review*, E M Purcell, H C Torrey (1911–1998) and R V Pound (1919–) in a short letter to the editor, received on 24 December 1945, announced that they had observed absorption of radio-frequency energy, due to transitions

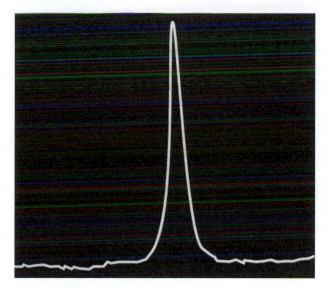

Figure 37. Absorption curve in the proton resonance of a ferric nitrate solution obtained with Purcell's method. (From Bloembergen N *et al* 1948 *Phys. Rev.* **73** 679).

induced between energy levels which corresponded to different orientations of the proton spin in a constant applied magnetic field in a solid material (paraffin).

In the experimental arrangement a paraffin sample was submitted to a static magnetic field in a resonant cavity. Radio-frequency power was introduced into the cavity at an extremely low level with its magnetic field everywhere perpendicular to the steady field, and its intensity at the output was measured. When the strong magnetic field was varied slowly, extremely sharp resonance absorption was observed (figure 37). Because the proton has spin-$\frac{1}{2}$, when it is subjected to a steady magnetic field, it may assume only two positions: either with its spin about parallel to the field or with the spin about anti-parallel to the field. The energy difference between the two energy levels that correspond to the two positions, with the strength of the magnetic field used in the experiment, corresponded to a frequency of 29.8 MHz. At this frequency the microwave field was absorbed (figure 37). At room temperature (at which the experiment was performed) the difference between the number of protons aligned with the magnetic field and the one anti-parallel to it was extremely small; however, the number of nuclei involved was so large that a measurable effect resulted once thermal equilibrium was reached. A crucial question concerned the time required for the establishment of thermal equilibrium between spins and lattice. A difference in the population of the two levels is a prerequisite for the observed absorption.

The authors kept this very clearly in mind and considered that competition existed between the absorption and stimulated emission processes. The absorption processes in fact involve the absorption of a photon from the microwave radiation, making the particle change from the lower to the higher level. The stimulated emission processes, on the contrary, correspond to the emission of a photon, which is similar to the stimulating photon, causing the particle to change from the higher to the lower energy level. Therefore if the two processes of emission and stimulated emission occurred in equal number, no signal would result. For this reason the establishment of thermal equilibrium is important because in this situation the lower energy levels are more numerous than those at higher energy, so that the absorption processes are preponderant.

Statistical mechanics may help. According to Boltzmann, in fact, the ratio between the number of molecules that are in a higher energy level and the number of molecules that are in the lower energy level is given by an exponential in which appears, with a negative sign, the energy difference between the two states divided by the factor kT, where k is a constant introduced by Boltzmann and T is the absolute temperature. In our case the energy difference between the two magnetic levels is proportional to the applied magnetic field, and thus increases when the field increases. However, for the values of the field achievable with standard techniques, this is always very small and is comparable with the value kT at room temperature. This fact means that, for example in the case of hydrogen, with a reasonable magnetic field (7000 Gauss) the difference in the populations of the lower and higher states respectively divided by the population in the lower state is almost one hundred thousandth at room temperature. This was, however, sufficient to detect the absorption signal.

Purcell's discovery may be considered the natural follow-on to the effort made during the war at the MIT Radiation Laboratory to decrease radar wavelengths to 1.25 cm. That wavelength happened to fall upon a strong absorption band of atmospheric water vapour and was therefore absorbed during its propagation in air, thus precluding practical radar operation. Purcell was interested in developing precise methods to measure absorption bands and in coherence with this he called his technique nuclear magnetic resonance absorption.

In the next issue of the *Physical Review*, again as a letter to the editor, there appeared a short note by F Bloch, W W Hansen (1909–1949) and M Packard (1921–), which had been received on 29 January 1946. The authors described an experiment, in some respect similar to the one of Purcell, obtained using water. In their experiment, over a constant magnetic field, applied along one direction (for example the vertical one), was superposed a small oscillating magnetic field directed along the horizontal direction. The nuclear magnetic moments of the substance, originally parallel to the constant field, are disturbed in this alignment by the small oscillating field and are forced to process around that field. At resonance frequency the small field may invert the direction of the magnetic moments and a signal may be picked up by a suitably positioned coil, through an effect of electro-magnetic induction already discovered by Faraday in 1832, according to which a variable magnetic field induces a current in an electric circuit.

In this research Bloch was motivated by the desire to find techniques to make more accurate magnetic field measurements. In 1946, Bloch also provided a theoretical explanation of the experiment by introducing two relaxation times for the spin population. One time describes how fast thermal equilibrium between nuclear spins and elastic vibrations of the material is reached (spin–lattice). The second time is the characteristic time with which the transverse component of magnetization relaxes to its equilibrium value which is zero.

Purcell and Bloch met for the first time at a meeting of the American Physical Society at Cambridge, in Massachusetts, in 1946. They always had cordial relations. When both were awarded the Nobel prize, Bloch sent to Purcell a telegram: 'I think it is splendid for Ed Purcell to divide the shock with Felix Bloch'.

Nuclear magnetic resonance, initially conceived as a means to study the magnetic behaviour of matter, with time has become an indispensable medical technique. Because it can measure the shift in the resonance frequency produced in the local nuclear environment, it has become a powerful method of chemical analysis that allows the identification of chemical compounds and the study of their structure. An important application is in medical diagnostics. The nuclear magnetic resonance allows in fact the position of nuclei-bearing magnetic moments to be identified as a result of the presence of their characteristic absorption spectrum. Nuclei that yield strong signals are for example hydrogen, deuterium, carbon and phosphorous. These nuclei can be identified by their nuclear resonance spectra and with special techniques their position in space can be found, so enabling three-dimensional images to be obtained.

The first spectra from living tissue were obtained only about 20 years ago. The reason why so much time was needed to develop this technique may be found by observing that in nuclear magnetic resonance (NMR), very small energies are associated with the transitions, and therefore to obtain sufficiently high signals very strong constant magnetic fields are required. These fields must, moreover, be extremely homogeneous over the region of interest which may be rather extended, like for example a human body. The use of superconducting magnets has overcome this difficulty. The medical applications of NMR today allow images of anatomic parts of the human body to be obtained and to identify chemical compounds in the organism. For these applications commercial devices exist whose use has entered the current practice of many large hospitals. The technique may replace the traditional use of x-rays, with improved sensitivity, and without producing the damaging collateral effects of the exposure to x-radiation. The 2003 Nobel prize in Physiology or Medicine was awarded to the American chemist Paul Lauterbur (1929–) at SUNY Stony Brook and the

British physicist Peter Mansfield (1933–) at the University of Nottingham 'for their contribution concerning magnetic resonance imaging'. In the 1970s the two researchers, independently from each other, set in motion the transformation of NMR technology from a spectroscopy laboratory discipline to a clinical imaging technology. The idea was that by changing the spatial value of the uniform magnetic field in an NMR spectrometer allows to localize from where the signal come. In fact, because the resonance frequency of a spin depends on the uniform magnetic strength if this strength is different in different points, the resonance frequency is different and therefore the knowledge of the value of the magnetic field in each point allows the localization of the nuclei whose transitions produce the NMR signal. In 1973, Lauterbur published the first spatially resolved image. In 1974–75, Mansfield and co-workers developed techniques for scanning samples rapidly and in 1976 obtained the first image of a living human body. The Nobel prize emphasized their contributions to speeding up the acquisition and display of localized images which was essential to the development of NMR as a useful clinical technique.

Whole body magnetic resonance imaging appeared on the scene about 1980. Compared with those vague images, today's pictures are spectacular. Details of organs with resolution of a few cubic millimetres that vividly display the organs can be obtained.

Electronic paramagnetic resonance

Electronic paramagnetic resonance does not differ substantially from nuclear resonance except for the fact that the energy levels created by the external magnetic field are not produced by nuclear spins, but are the Zeeman levels produced by the effect of the magnetic field on the motion of electrons in the atom. As we have already seen, the application of an external magnetic field to an atom removes the energy degeneration of the atomic orbits, and each electronic energy level is split into many sub-levels separated by a small amount of energy which typically corresponds to microwave frequencies, while the separation among the levels depends on the external magnetic field.

The principle of the method is very simple. A constant magnetic field is applied to the substance so that the electronic levels experience Zeeman splitting. Simultaneously a small radio-frequency field is incident and its frequency is varied. An absorption peak results when the frequency of the variable field corresponds exactly to the energy separation between two Zeeman levels. An example may be discussed by making reference to chromium ions present in ruby (later on we will discuss ruby in detail). These ions have unpaired electrons that contribute to a total spin equal to 3/2. The electric field, created by all the atoms present in the crystal, blocks all the other angular moments and therefore in the presence of an external magnetic field the behaviour arises only from the unpaired electrons. The diagram of the created levels depends on the orientation of the external magnetic field with respect to the principal symmetry axis of the crystal (figure 38a). This effect is a notable one, as can be seen from figures 38b

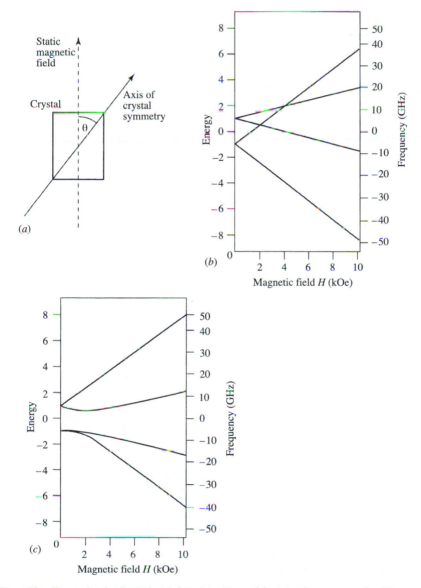

Figure 38. Energy levels of ruby in (a) the orientations of the crystal symmetry axis with respect to the direction of the magnetic field. The energy levels of ruby for (b) $\theta = 0°$ and (c) $\theta = 90°$. (From Thorp J S 1967 *Masers and Lasers: Physics and Design* (London: Macmillan).

and 38c that depict the energy levels as a function of the strength of the magnetic field for two different orientations. It is worth observing that by choosing a suitable value for the external magnetic field, the desired separation between the levels can be obtained. This is the fundamental operating principle of the three-level maser.

Atomic clocks

In 1949 N Ramsey invented the resonance technique with separated oscillatory fields of which we have already spoken, and which in 1955 was applied by Zacharias, J V Parry, Louis Essen and others to build atomic clocks and frequency standards. For this method, Ramsey was awarded the physics Nobel prize in 1989 together with H Dehmelt (1922–) and W Paul (1913–1993) who developed sophisticated techniques to study isolated atoms or molecules.

The problem of time measurement has always been important and difficult. At the beginning it was linked to the rotation of the Earth around itself, which was supposed to occur with great regularity. The increasing accuracy of pendula introduced by Huygens and of astronomic observation, in Newton's day, led John Flamsteed (1646–1719), the first Astronomer Royal at Greenwich, to test the regularity of the Earth rotation, using pendulum clocks. Flamsteed did not find any evidence that the Earth's rotation was less than perfectly uniform, but successive generations of astronomers with time have compiled a lengthening list of irregularities in the length of the day.

Early in the 20th century, for example, through astronomical observations, the slowing down of the Earth due to tidal friction was definitively accepted. By the mid 1930s, the pendulum clock had improved to the extent that it allowed the irregularities of the Earth's motion to be measured (figure 39). At this time the quartz clock entered the stage and measured with more accuracy these irregularities. In quartz clocks the vibrations of a quartz crystal are triggered electrically and used to generate a constant frequency that strikes the time. A quartz clock may be calibrated through astronomical observation and used to make time measurements in the laboratory. The best one presently may work for one year accumulating a maximum error of 5 millionths of a second. This precision is, however, still insufficient for scientific and technological purposes.

Once the irregularities of the Earth's motion were recognized, it was necessary to find another way to define the time standard, which until then was the second defined as the 86 400th part of the mean solar day, obtained by astronomers by considering the motion of the Earth around itself during the whole solar year. Astronomers were disinclined to lose their primacy in defining the time and proposed to link the second to the annual revolution of the Earth around the sun (Ephemeris Time), taking as a reference the

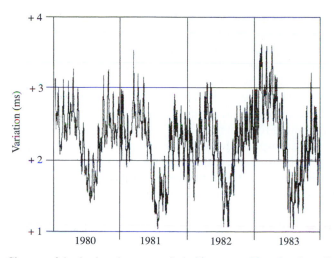

Figure 39. Changes of the day length over a period of four years. Note that the ordinate scale is only 3 ms = 0.003 s wide. (From Wahr J 1985 *American Scientist* January–February p 41).

mean year of 1900. This measurement unit was adopted by the General Conference of Weights and Measures of 1960 (one second was defined as 1/31 556 925.9747 of the year 1900).

However stable a pendulum or a quartz clock may be, they were no more able to satisfy one of the most important criteria, namely independent reproducibility.

The development of atomic theory and, in particular, the conclusion that atoms of a given chemical element were identical to one another suggested as a basis for the definition of the units of measurement not the Earth, but the atom itself. Already Maxwell and Lord Kelvin had suggested using, as a unit of measurement of the length and time, the wavelength and the corresponding frequency of some radiation emitted by a suitable atom, for example hydrogen, in the simplest case, or the sodium D line which is one of the most intense lines. Several years were needed, however, before this idea found practical application. Just after the Second World War, C H Townes at Bell Laboratories and R V Pound at the MIT Radiation Laboratory suggested using microwave absorption to stabilize an oscillator.

If we consider an oscillator that emits microwaves, for a great number of reasons its frequency does not remain rigorously stable in time, but suffers small casual variations. The problem, if the frequency of these oscillations is to be used, is therefore to find a way to keep it constant. The solution proposed by Townes and Pound was to use an ammonia molecule which has an absorption that is a function of frequency with a maximum exactly at 23.8 GHz which does not change in time for any reason. The principle is very simple. Let us examine it in the specific case of ammonia. Assume we direct microwave radiation at a frequency that can be varied around

24 GHz into a microwave cavity filled with ammonia, and measure the output power, that is the absorption by the ammonia. By varying the frequency the absorption is maximum just at the central frequency of the ammonia line (23.8 GHz). When this absorption is maximum we know that the frequency of the microwaves we are sending into the cavity is just 23.8 GHz, and therefore we have achieved a very precise frequency. We may fix the parameters of the oscillator in such a way that it oscillates exactly at this frequency. If for any reason the oscillator suffers a small change in its frequency, the absorption decreases and with a suitable feedback system we may re-adjust the frequency to its proper value. The method allows us to control and fix constant the microwave frequency over long periods of time using an absorption process: in this case the absorption at 23.8 GHz of the ammonia molecule.

In the years 1947–48, Townes at Bell Laboratories and three other researchers, one at RCA and two others in two American Universities, succeeded in assembling four similar devices that stabilized the frequency emitted by a klystron using ammonia. All these devices were patented. To have a clock, however, one needs to divide the obtained high frequency at least by a factor of one thousand so as to bring it into the megahertz range where it may be contrasted with suitable oscillators that in this way are stabilized, and then drive the clocks.

Harold Lyons (1913–1991), responsible for the division for microwave standards at the American Bureau of Standards, pushed by Townes, in August 1948 built a frequency standard which was stabilized using the ammonia transition. In 1952 his group successfully obtained a stability of one or two parts per hundred million; however, this was not much better than that of the Earth's rotation. Strenuous efforts by K Shimoda at the University of Tokyo succeeded in improving this stability to one or two parts in 10^9.

Already in 1948, when he was still assembling his first frequency standard using ammonia absorption, Lyons had also started a programme for assembling a true clock that utilized atomic transitions, that is an 'atomic clock'. Because he was not an expert in spectroscopy he asked the assistance of Isidor Rabi. Rabi's right hand man, Polykarp Kush, prepared a conceptual design in which the use of a caesium atomic beam was envisaged. Caesium had already been extensively studied by the Rabi group and for a number of reasons promised to produce the highest possible stability.

Polykarp Kush (1911–1993) was a very capable experimentalist. After his PhD in molecular optical spectroscopy, obtained in 1936 with a thesis under the supervision of F Wheeler Loomis at Illinois University, in 1937 he went to Columbia where he collaborated with Rabi on pioneering the method of magnetic resonance of molecular beams. With this technique he made a number of measurements among which were those which led him to the discovery of the anomalous magnetic moment of the electron in 1947 for which he was awarded the physics Nobel prize in 1955 together with Willis E Lamb. During the Second World War he worked on the development of high frequency generators for radar. His lectures, characterized by a very sonorous voice (known as the 'Kush whisper'), had the quality of a sermon, a trait which probably he had inherited from his father, a Lutheran missionary.

In the summer of 1951, the apparatus started to operate, although not in the final configuration of the design. Kush in fact had sought to employ Ramsey's two-fields technique to accurately tune the frequency of the microwave field. This was the first experiment using this technique. In the spring of 1952, the device worked with this method and by the end of the year the transition frequency of the isotope of mass 133 of caesium, the only one that is found to be stable in nature, was measured and found to be 9.192 631 800 GHz.

These positive results also pushed Britain's National Physical Laboratory to build its own version of the device, still based on the same technique of Ramsey's double resonance. By contrasting their measurements with accurate astronomical measurements, performed at Greenwich, Luis Essem and J V L Parry established the radiation frequency more exactly as 9.192 631 770 GHz which was adopted in 1964 by the General Conference of Weights and Measures and since then has been accepted as the official definition of the second (that is in one second there are 9 192 631 770 oscillations of the caesium atom). This definition made atomic time agree with the second based on Ephemeris Time.

All the devices described until now were not proper clocks, in the sense that they did not allow a direct comparison with the much lower frequencies that could be used to count seconds. The final step was performed at the MIT Radiation Laboratory by Jerrold Zacharias (1905–1986) who, in 1955, succeeded in building a version of an atomic clock, still using caesium, that was to be commercialized under the name of 'atomichron'. In 1967, the international time standard for the second was defined in terms of the hyperfine frequency of the caesium atom.

The stability and accuracy of atomic clocks have experienced a decisive advance in the 1990s through the development of trapping and cooling atoms by means of lasers. Complex geometries have permitted to built what are called fountain clocks. The duration of the atom-wave interaction in an atomic clock is finite and broadens the resonance peak as a result of the Heisenberg time–energy uncertainty principle. Other effects affect the frequency limiting the accuracy of the clock. Atoms can now be cooled relatively easily down to $1\,\mu K$. At such temperatures the atoms have a thermal velocity of only a few millimetres per second, instead of the 100 m/s they have at room temperature.

In the so-called atomic fountain the atoms are launched vertically upwards using lasers and eventually fall back down again due to gravity, like water in a fountain. The system is so designed that atoms interact with the electromagnetic field both on the way up and the way down so increasing the interaction time. With a height of 1 m the interaction time in current atomic fountains approaches 1 s.

The idea of a fountain clock was already considered in 1953 by Jerrod Zacharias, but collisions between atoms in the beam did not allow success.

In 1989 Steven Chu and Kurt Gibble at the Stanford University (USA) demonstrated the first fountain using cooled sodium atoms, followed in 1991 by Claude Cohen-Tannoudji and Christophe Salomon at the Ecole Normale Supérieure in Paris together with André Clairon at the Paris Observatory using caesium. In 1993 Clairon built the first cold-atom fountain which in 1995 achieved an accuracy of less than 1 s in 30 million years (an enormous accomplishment that scientists are already studying how to improve.)

The need to have such accurate clocks occurs for example in radio astronomy or to verify Einstein's relativity theory. The most prominent use of atomic clocks is perhaps in the global positioning system (GPS) of satellites for navigation and monitoring. Each of the 24 satellites in the network emits a signal that identifies its position at that time, which is referenced to the atomic clock that is on board. The distance between a person holding a receiver and the satellite is derived from the time difference between the emission and reception of the signal. At present, GPS has a precision of a few metres, which requires time measurements that are accurate within a few nanoseconds (10^{-9} s).

Experimental proof of population inversion

Returning to nuclear resonance, certain experiments undertaken by Bloch and co-workers in 1946 are of great relevance to our story. In some measurements they had found that the relaxation time of water was between half a second and one minute. To determine its value more precisely an ingenious experiment was designed. A sample of water was submitted to a static magnetic field of sufficiently high strength. By applying the oscillating field and by changing its frequency, the typical resonance peak was obtained. Thereupon, in a very short time the direction of the static field along which the magnetic moments precessed was inverted. Initially, the resonance peak was observed, then in the following few seconds the peak disappeared and grew again with negative values. These findings may be explained in the following way. Initially, the magnetic moments are aligned along the field direction and the weak alternating field through absorption induces transitions towards the higher energy level (this is the origin of the absorption peak). When the static field changes direction, initially the spins are oriented nearly anti-parallel to the field and therefore the alternating field induces now stimulated transitions from higher to lower energy levels until slowly the system goes into equilibrium in this new direction.

In Bloch's experiment the time needed to reach the new equilibrium was around a few seconds and this was also the value of the relaxation time of water in which he was interested. During the few seconds necessary to re-establish equilibrium, however, the spin population was in a condition of population inversion, that is with more particles in the magnetic level of higher energy than in that of lower energy.

Bloch paid no attention to this, concentrating instead on the problems of determining the relaxation time, its exact meaning and its value. The inversion of population obtained in this way—which takes the name of adiabatic fast passage—was later used in 1958 to create population inversion in two-level solid-state masers.

The following year, N Bloembergen, a young Dutch physicist, of whom we shall speak in detail later, together with Purcell and Pound, continued the study of relaxation times, introducing into the theoretical discussion the equations that govern the behaviour of the number of atoms residing at different energy levels, which were to prove fundamental to the description of the behaviour of masers and lasers.

Negative temperature

In 1951, Purcell and Pound, in a very short note in the *Physical Review* introduced the concept of negative temperature and showed the existence of negative absorption. They considered a nuclear absorption experiment and reasoned in the following manner. At field strengths which allow the system to be described by its net magnetic moment (magnetization), a sufficiently rapid reversal of the magnetic field should result in a magnetization opposed to the new direction of the field. The reversal must occur in such a short time that the magnetization is not able to follow the field instantaneously. They performed an experiment with a lithium fluoride sample, placing it in a magnetic field whose direction, after equilibrium was attained, was suddenly inverted. The inversion time was made shorter than the spin–lattice relaxation time, and so the configuration of nuclear spins had insufficient time to change during the field inversion. During the short time in which spins stayed inverted, a negative absorption (i.e. an emission) occurred.

This effect is shown in figure 40, which is one of the records obtained by sweeping the impressed frequency periodically back and forth through the resonance frequency. The peak at the extreme left is the normal resonance curve, before the field is reversed. Just to the right of this sweep the field has been reversed and the next resonance peak is seen to point downwards, corresponding to negative absorption. The negative peaks get weaker until finally the state is reached where the positive and negative absorption cancel out because there is then equal population of the upper and lower states. The gradually increasing positive peaks show the re-establishment of the thermodynamic equilibrium population.

If we now remember that if we have molecules that may reside in two energy levels, the number of molecules that are in the upper state is equal to

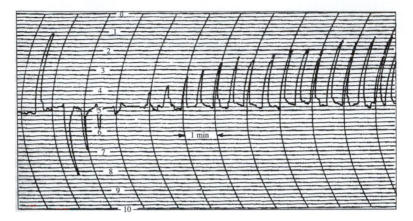

Figure 40. Typical registration of the nuclear magnetization reversal. (From Purcell E M and Pound R V 1951 Physical Review **81** 279).

the number of molecules that are in the lower state multiplied by an exponential factor in which appears the negative of the ratio of the energy difference between the two levels divided by kT, we see that for a positive value of the temperature, the argument of the exponential function is always negative and therefore the exponential term is always less than unity. This means that the population of particles in the upper state is always less than that in the lower energy level. The situation in which more particles are in the upper level with respect to those in the lower level corresponds to the case in which the exponential is greater than unity, which may be obtained if its argument is positive. This corresponds to a negative temperature.

A negative temperature simply means that the occupancy of the higher energy states is more probable than the lower energy state; it is a useful way to treat the case in which population inversion occurs using the same equations. However, one must not think that this is only a convenient way of making calculations: it bears a true physical meaning. Seven years after its introduction, a full explanation of its meaning was given by two researchers, A Abragam (1914–) and Proctor. Purcell commented on the paper saying: 'It is like receiving a marriage licence seven years after the child is born'.

N Ramsey, who discussed in several papers the (even philosophical) meaning of negative temperature wrote:

> 'Pound, Purcell and Ramsey performed a series of experiments with
> LiF crystals which have a very long relaxation time. They found,
> among other things, that the spin system is essentially isolated for
> times which vary from 15 s to 5 min and that, for times short
> compared to these, the spin system can be placed in a state of negative
> temperature. In a negative temperature state the high-energy levels are

occupied more fully than the low, and the system has the characteristic
that, when radiation is applied to it, stimulated emission exceeds
absorption.'

In the Purcell and Pound experiment, the signal they observed was produced
by the decay of the inverted population in Zeeman levels. Nobody paid any
attention to this method which allowed inversion to be obtained or to the fact
that systems at negative temperature, when in connection with a microwave
cavity or a waveguide, could yield coherent amplification through stimulated
emission processes. This was probably due to the fact that the method of
attaining inversion produced only transient population inversions. Only
after its invention were these methods used for making masers.

CHAPTER 10

THE MASER

Now we are ready to recount how the maser was invented, or better to describe how it was discovered, considering that in the universe stellar maser sources exist and therefore man has effectively reproduced in the laboratory what already existed in nature.

To gain an exhaustive picture of the environment in which first the maser and then the laser developed, it is useful to summarize the status of research in physics and the spirit with which this research was carried out by physicists in the years after the Second World War. Before the First World War, studies were performed mainly by individuals or small groups who had no connection with industry and did not look towards any particular application. Scientists strived to satisfy their own curiosity to discover and explain new phenomena. The discovery of radium by Marie Curie (1867–1934) is a clear example. She started with the observation that some minerals from which uranium, the radioactive element just discovered in 1896 by Henri Becquerel (1852–1908), was extracted, showed an activity much larger than could be attributed to the uranium content and that, therefore, within the mineral one or more substances much more radioactive than uranium must exist. This pushed her to investigate these substances. The research was performed in a shed, with many difficulties, without the help of anybody except her husband Pierre (1859–1906). The final result was the discovery of polonium and radium (1898). It was soon found that the radiation emitted by radioactive substances had biological effects and they were thus used in the struggle against cancers. The discovery of radium therefore led to an important practical application that nobody could have foreseen at the beginning of the research.

During the Second World War, the situation changed radically, especially in the United States, where the most important results of our story were obtained. The development of radar, which massively aided the war effort, and nuclear research, culminating with the construction of the bomb, showed what enormous potential existed in physical research and what extraordinary applications could be obtained. At the end of the war, physics presented itself as a science for which physicists were the priests capable of providing mankind with useful services.

176

In 1948 the invention of the transistor and the consequent revolution in the field of electronics provided the decisive step-change. Now research was no more an end to itself with the sole scope of extending knowledge in general, but became a means of acquiring new knowledge leading to concrete applications of notable social impact.

Physics was discovered not to be an abstract science for a few of the initiated, but an instrument able to provide basic elements for the development of society or, depending on the approach that one was willing to adopt, for its destruction. Nuclear research and the development of nuclear reactors for energy production, today regarded with suspicion, at the end of World War II were looked upon with great favour as a means of solving energy problems and thus benefit mankind.

At this time began a link with industry, and the creation of industrial research laboratories and industries based on physics research ensued. Suddenly, physicists became important and popular and a huge amount of money was given to them from government sources. In this atmosphere, anybody who had a good idea that potentially could lead to applications had the quasi-certainty of finding adequate support to develop it. This ideal environment lasted for the entire wartime period and continued during the Cold War between the United States and Russia until the 1970s. In this period the development of masers and lasers was helped in America through the interest of both military agencies and industry, which were receiving financial support at a level never seen before. In other countries, this happy state of research and its executors never reached such levels although some improvement was certainly attained.

In the ex-Soviet Union, the Academy of Sciences could be considered a state within the State, very rich and powerful. In its institutes distributed throughout the country, thousands of researchers were working for much higher than average earnings compared with other workers. There also existed secret laboratories for military research which, together with the excellent laboratories of the Academy, were the only ones financially capable of importing the necessary instrumentation.

In Europe large international research enterprises were founded such as CERN in Geneva. This large laboratory was established with the financial support of Italy, France, Germany, Great Britain, the Netherlands and Belgium to build powerful particle accelerators for research into high energy physics using facilities whose funding was impossible for a single country. The Geneva accelerators provided for the numerous and well organized research groups a great number of high energy particles. In the years before and during the war, experiments were laboriously performed by waiting for the spontaneous arrival of the very small number of high energy particles present in cosmic rays, their natural source. The development of the accelerators meant that these particles and their interactions could be obtained in great number through trivial routine work.

Large financial support was given in Europe and especially in Italy for research into high energy particles where competence in the field came from the roots laid down by the prestigious Fermi, Amaldi, Persico and others. Such funding was directed in a well defined direction.

The fact that the development of the new devices, the maser and the laser, took place in the United States and in the ex-Soviet Union, the sole countries in which research encompassed a broad range of fields, can be attributed to the fundamental strategy of not focusing research into specific directions.

The great change in the role of physics and of physicists in research that took place during the Second World War needs to be considered to understand how costly research with unguaranteed and often doubtful outcome could be supported and carried out without great problem. The maser implies an amplification technique so radically different from usual techniques that it could not originate as a simple improvement of the electronic techniques already known, but required the development of new fields such as magnetic resonance and microwave spectroscopy, that occurred principally in the United States and which we described in previous chapters. Moreover until then electromagnetic radiation was produced and detected using tubes (diodes, triodes etc.) which today are completely obsolete, that operated using electrons emitted by a filament heated by an electric current. The action of these devices, as well as of magnetrons, klystrons etc., was perfectly understandable by applying Maxwell's laws of classical electromagnetism. Engineers had no need to study quantum mechanics, which is indispensable in understanding the operation of the maser, until 1948 when the transistor was discovered. To understand how this device worked it was necessary to consider the electronic states in solids which are described by quantum mechanical laws. At this point engineers discovered quantum mechanics and started to study it.

Immediately after the war, the work on microwave spectroscopy in America was initially mainly developed in industrial labs. Four independent groups started the study of gases by means of these techniques: Bell Telephone Laboratories, Westinghouse, the RCA Laboratories and Columbia University as the sole university representative. After a few years this activity, considered to be of low profitability, ceased in industrial laboratories and moved to the universities where physicists and chemists were interested in utilizing microwaves for the study of molecules. However, most molecules have the most intense spectral lines at wavelengths in the millimetre region and interact very weakly with radiation of wavelength around a centimetre, produced by radar sources. This provided a strong incentive to molecular spectroscopists to develop microwave sources in the millimetre and submillimetre regions. Military personnel were also interested in millimetre systems which were more compact and lighter for radar instrumentation and for other wartime applications.

During the war all the most important American scientists had collaborated with the military and, now that the war was over, the collaborations continued naturally, even if on different projects, favoured by reciprocal personal acquaintances. Within this atmosphere, in 1946 the Office of Naval Research, the Army Signal Corps and the Army Air Force created the Joint Services Electronics Program (JSEP) to support the two laboratories which during the war had performed electronics research and which now in peacetime were still maintained: the MIT Radiation Laboratory, reorganized into the Research Laboratory of Electronics, and the Columbia Radiation Laboratory of the Physics Department of Columbia University. At the end of 1946, a JSEP program also started at Harvard University and in 1947 the Stanford Electronics Laboratory in California was included.

The American Defense Department had high regard for the JSEP laboratories that promoted science and technology from which one could expect advances in military areas and moreover helped to update important scientists on military requirements and to form future generations of researchers. So researchers and military personnel were working with a strong mutual interaction in the field of millimetre wave generation.

In the 1950s, the time was ripe for the invention of the new device and as often happens the fundamental and working principles were provided contemporarily and independently by several different men.

The reader who has not been distracted by the numerous digressions indispensable for understanding our story should see a precise path that brings us to the final application. After a tortuous beginning, the journey progresses with the introduction by Einstein in 1916 of the concept of stimulated emission, which in the 1920s was first used by theoreticians to explain various phenomena and which was later experimentally verified. While the concept gradually settled in the minds of researchers, in the 1930s microwaves and related techniques exploded and immediately after intense research carried out on radar during the war, magnetic resonance was discovered. In magnetic resonance experiments, the first population inversions were obtained; timid, transient but real and correctly identified as such. Also identified was the role of stimulated emission and its contraposition to the phenomenon of absorption. Let us now see how stimulated emission was applied to the creation of the maser.

The Weber proposal

The first description in a public audience of the basic principle according to which a maser may work (without, however, a working device) was given by Joseph Weber (1919–2000) in Ottawa, Canada, in 1952 during a conference on electronic tubes—the Electron Tube Research Conference, a prestigious conference to which participation was only by invitation and at which new ideas for advanced devices were often presented.

Weber was then a young electrical engineering professor at the University of Maryland and a consultant at the United States Naval Ordinance Laboratory. He graduated in Annapolis and was first a naval officer from 1945 to 1948 having responsibility, as a specialized microwave engineer, for the section on electronic countermeasures of the Navy. Here he had the opportunity to become acquainted with the technological importance of amplifiers with high sensitivity at microwave and millimetre wavelengths, since the receivers that are required for countermeasures against enemy radar waves employ such amplifiers to detect very faint radar waves. Information about the employed wavelength and its origin is then utilized to send signals that blind the enemy receivers, hindering their identification of targets.

He resigned from the Navy and entered graduate school at the Catholic University of Washington. Quantum mechanics was part of his studies and the idea of the maser came after attending a lecture on stimulated emission give by Karl Herzfeld (1892–1978). He received his PhD in physics in 1951 working in microwave spectroscopy and then joined the University of Maryland as a full professor of electrical engineering. Here he continued to work on microwave spectroscopy. The mechanism of absorption and emission that takes place when a gas interacts with radiation had always interested him. In a typical microwave spectroscopy experiment, microwaves produced by some source impinge on a detector. If a gas is inserted between source and detector, absorption of some of the incident radiation can be observed. What is the nature of this absorption? It occurs if the gas molecules possess a pair of levels whose energy difference divided by Planck's constant is about equal to the microwave frequency. To better understand what happens, Weber considered a system with only two energy levels, E_1 and E_2 (with E_2 greater than E_1) in each of which there is a number n_1 and n_2 respectively of atoms or molecules (we will call n_1 and n_2 the populations of the energy levels E_1 and E_2 respectively). When the microwave frequency has the right value, the power absorbed is proportional to the population of the first level, that is n_1. The particles that are in the upper level 2, for their part, emit by stimulated emission at the same frequency and the power of this emission is proportional to n_2. The net power is equal to the difference between the absorbed and emitted powers, i.e. is proportional to $n_1 - n_2$. Because at thermal equilibrium n_1 is always larger than n_2, Weber observed that 'this net power is always a positive quantity. Thus under ordinary circumstances we get absorption of radiation'. However, he added 'We could get amplification if somehow the number of oscillators in the upper states could be made greater than the number in the lower states' and concluded 'A method of doing this is suggested by Purcell's negative temperature experiment'.

These considerations were developed by Weber in 1951 and presented at the conference in 1952. He published them in summary form in 1953. As

Weber explained later, it was his intention to publish his results in a widely read journal. Instead early in 1953, Professor H J Reich of Yale University wrote to him saying that he had been chairman of the 1952 conference programme committee, and was also editor of a (not so widely read, according to Weber) journal and was proposing to publish the conference papers in it. As a result the conference summary report was published in the June 1953 issue of *Transaction of the Institute of Radio Engineers Professional Group on Electron Devices*.

In his work Weber underlined the fact that the amplification is coherent. The method he proposed for obtaining population inversion has never, in fact, been put into practice and it seems most unlikely ever to be so. Moreover Weber was only interested in an amplifier. The idea of feedback, so essential in Townes' maser, as we will see, was not important for Weber and thus he did not discuss it. Weber evaluated also quantitatively the possibilities of his device but the numbers he obtained in calculations showed such minimal performances that he decided to give up with it and not try to build anything. However, the idea aroused some interest, and after presentation of his work at the conference, Weber was asked by RCA to give a seminar on his idea. For this he received a fee of $50. After the seminar Townes wrote to him, asking for a reprint of the paper. Weber's work was, however, not quoted in the first of Townes' papers but was referred to later.

Weber's efforts were acknowledged by the IRE when he was awarded a fellowship in 1958 'for his early recognition of concepts leading to the maser'. He spent the 1955–56 academic year as a fellow of the Institute for Advanced Study in Princeton, and immersed himself in general relativity. During the early 1960s he was interested in gravitational waves, building detectors which, however, failed to detect them definitively.

Townes and the first maser

The most well known work that eventually produced an operating device was carried out at Columbia University in which considerable research into radio frequency spectroscopy was performed, supported by a far-sighted military contract with the Armed Services. The result was that a group of researchers, headed by C H Townes, built and operated the first maser.

Charles H Townes was born in 1915 in Greenville, South Carolina. When he was only 16 he entered Furman University. Although he soon discovered his vocation for physics, he also studied Greek, Latin, Anglo-Saxon, French and German, and received a BA degree in modern languages after three years at Furman. At the end of his fourth year he received a BSc in physics. Next he attended Duke University on a scholarship. When he was 21 he finished work on his master's degree, continuing to study French, Russian and Italian. He then went to the California Institute of Technology in Pasadena, where, in 1939, he received his PhD, after which he accepted an appointment at Bell Telephone Laboratories. During the war Townes was assigned to work with Dean Wooldridge, who was then designing radar bombing systems. Although Townes preferred theoretical physics, he nevertheless worked on this practical project.

At that time people were trying to push the operational frequency of radar higher. The Air Force asked Bell to design a radar at 2400 MHz. Such a radar would exploit an almost unexplored frequency range and would result in more precise bombing equipment.

Townes, however, having read an unpublished memorandum by van Vleck on the theory of atmospheric water vapour absorption, observed that radiation of that frequency is strongly absorbed by water vapour. The Air Force nevertheless insisted on trying it. So the radar was built by Townes, who then was able to verify that it did not work. As a result of this work, Townes became interested in microwave spectroscopy.

In 1947 Townes accepted an invitation from Isidor I Rabi to leave Bell Laboratories and join the faculty at Columbia University in which Rabi was working. The Radiation Laboratory group in the Physics Department there had continued the war-time programme on magnetrons for the generation of millimetre waves. This laboratory was supported by a Joint Services contract from the US Army, Navy and Air Force, with the general purpose of exploring the microwave region and extending it to shorter wavelengths. Among the people active in the sponsorship of this programme were Dr Harold Zahl (1905–1973) of the Army Signal Corps and Paul S Johnson of the Naval Office of Research. Townes quickly became an authority on microwave spectroscopy and on the use of microwaves for the study of matter. In those years Townes was interested in the building of atomic clocks using the microwave absorption of ammonia to stabilize the frequency.

In 1950 he became full professor of physics. In the same year Johnson organized a study commission on millimetre waves and asked Townes to take the chair. Townes worked on the committee for nearly two years and became rather dissatisfied with its progress. Then, one day in the spring of 1951, when he was in Washington, DC, to attend a meeting of the committee, he tells us:

'By coincidence, I was in a hotel room with my friend and colleague Arthur L Shawlow, later to be involved with the laser. I awoke early in the morning and, in order not to disturb him, went out and sat on a nearby park bench to puzzle over what was the essential reason we had failed [in producing a millimetre wave generator]. It was clear that what was needed was a way of making a very small, precise resonator and having in it some form of energy which could be coupled to an electromagnetic field. But that was a description of a molecule, and the technical difficulty for man to make such small resonators and provide energy meant that any real hope had to be based on finding a way of using molecules! Perhaps it was the fresh morning air that made me

suddenly see that this was possible: in a few minutes I sketched out
and calculated requirements for a molecular-beam system to separate
high-energy molecules from lower ones and send them through a
cavity which would contain the electromagnetic radiation to stimulate
further emission from the molecules, thus providing feedback and
continuous oscillation.'

He thought that there would be only a slim chance of success; therefore he
did not say anything at the meeting and once back in Columbia in the
autumn of 1951 when Jim Gordon came to him, seeking a thesis project,
he started work. In addition to Gordon, Herb Zeiger was asked to join the
project because Townes reasoned that someone expert in molecular beam
work, on which he had just completed a thesis, would be helpful. Zeiger
was supported by a scholarship from the Union Carbide Corporation. The
chemical engineer H W Schultz, with a prophetic spirit, a couple of years
before the work on the maser started, had persuaded his company, the
Union Carbide, to give $10 000 per year to someone to study how to
create intense infrared radiation, having understood its importance to
induce specific chemical reactions. Although Townes insisted he did not
know how to solve the problem, even if he was very interested in it, Schulz
gave him the money for a post-doctorate assistant and that money was
used to support Shawlow and Herbert Zeiger in the years before the work
on the maser started. It was such that Zeiger was able to join the project proposed by Townes.

In Townes' design, the resonator was very important. In fact the cavity
was required to confine the electromagnetic energy for the longest possible
time so as to interact with the molecules (that is it should not have significantly high losses). Detailed calculations performed in the autumn of 1951
showed it was very difficult to make a cavity for radiation of half a millimetre
as Townes had initially thought would be achievable using molecules of
ammonia deuterate. Therefore he decided to turn his attention to radiation
of 1.25 cm, emitted by ordinary ammonia, which was the wavelength at
which the components necessary for success (the cavity) already existed.
The decision was therefore made to switch from a project that had to advance
the frontiers of research in the millimetre region to one that demonstrated a
new principle for obtaining generation in an already known spectral region.

The basic idea was very simple and, now that we are aware of it, we may
be amazed that nobody had thought at it before. If we consider a system with
two energy levels, as the one considered by Weber, the power emitted by
stimulated emission is proportional to the number n_2 of particles in the
upper state. The absorbed power is instead proportional to the number n_1
of particles in the lower state. The net power, that is the difference between
the absorbed and emitted power is, as we saw, proportional to the difference
$n_1 - n_2$. At equilibrium, n_1 is always larger than n_2 and therefore the
absorbed power is always larger than the emitted power. But let us consider

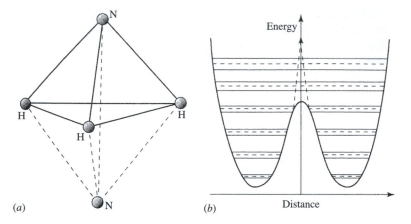

Figure 41. (a) Pyramidal structure of the ammonia molecule. (b) Potential energy of the nitrogen atom as a function of its distance from the hydrogen atom plane.

what would happen if in some way we could select the particles so as to send to one side only those in the upper state. The number n_2 is now larger than n_1 and therefore the emitted power is larger than the absorbed power. We have made a device able to emit the radiation corresponding to the energy difference between the two levels. Such a device is an oscillator and it was just this about which Townes was thinking.

The active material envisaged by Townes was ammonia gas. In the classical picture, the ammonia molecule (composed of one atom of nitrogen and three atoms of hydrogen, NH_3) is structured like a triangular pyramid (figure 41a) with the three hydrogen atoms at the vertices of the base and the nitrogen atom at the apex. The three hydrogen atoms can be considered to lie in a plane and the nitrogen atom to be above or below this plane. The potential energy of nitrogen as a function of its distance from the plane of the hydrogen atoms is shown in figure 41b. Quantum mechanics shows that it has a minimum at both sides of the plane with an energy barrier that has its maximum in the plane of the hydrogen atom. The nitrogen atom oscillates along an axis perpendicular to the hydrogen plane and may pass from a position above to one below the plane and vice-versa. Such a transition is described as inversion. Moreover the molecule rotates around two mutually orthogonal axes. According to quantum mechanics, all vibrational and rotational motions are quantized and therefore their energies are represented by discrete energy levels as shown in figure 41b.

Gordon, Zeiger and Townes, after some modification of their ideas, decided to look at the transition between the lower vibrational pair in the rotational state with three units of angular momentum about each axis, called the 3–3 state. This transition corresponds to a frequency of 23 830 MHz.

Townes' idea was to produce in some way a beam of molecules, for example by heating, and then to separate those that were in the excited state from those in the lower state. This could be done by considering an interesting property of the molecule: by applying an electric field the molecule is deformed and a small electric dipole moment appears in both members of the rotational pair, but opposite in sign in each. If the electric field is inhomogeneous, the same effect seen in the molecular beams of Stern and Gerlach occurs, and a force will act on the molecule which has opposite direction for each member of the pair. The project was then to use a strong electrostatic field to obtain a beam of excited ammonia molecules and to focus this beam through a small hole in a box, that is a resonant cavity, which had been tuned to resonance at exactly 23 830 MHz.

In the 1950s, Wolfang Paul (1913–1993) together with Helmut Friedburg and Hans Gerd Bennewitz designed special electric and magnetic lenses (quadrupoles and exapoles) to focus atomic and molecular beams. Townes thought about using lenses of this kind to separate the molecules. The fields created by these lenses were hence used in the ammonia maser and then in the hydrogen maser. Later, Paul developed a three-dimensional version capable of confining ions in a small region (the Paul trap) and for this achievement, which allows the behaviour of a single atom to be studied, previously impossible, he was awarded the Nobel prize in physics in 1989 together with N Ramsey and Hans Dehmelt who built a similar trap but of different design.

Townes was hoping to be able to select in the beam more molecules in the upper state than those in the fundamental state. In this way he could realize what we have called a population inversion and each molecule, on decaying into its fundamental state, would trigger other molecules to do the same. Therefore the cavity would emit coherent radiation at about 24 000 MHz. It is to the credit of Townes that he clearly understood at the outset the need for a resonant cavity with which to couple the radiation to the excited medium.

He felt responsible, particularly for Gordon who had to carry out his doctorate work on a project for which the result was not exactly known. 'I'm not sure it will work, but there are other things we can do with it if it doesn't' and so Townes promised Gordon that if the method did not work he could use the set-up to investigate the microwave absorption spectrum of ammonia. Gordon thus worked simultaneously on both experiments. In this manner he was able to study the hyperfine structure of ammonia (the separation of the energy states in many sub-levels due to the interaction between the electron and the nucleus) with an accuracy higher than had been possible before.

Progress in the work was described in quarterly reports of the Department and was given a certain amount of circulation among scientists who were interested in microwave physics. The first published mention of this maser project appeared, under the names of Zeiger and Gordon, in a report of 31 December 1951, headed *Molecular Beam Oscillator*. Preliminary calculations of the essential elements of the oscillator were reported there.

For two years the Townes group worked on. At about this time two friends called at the laboratory and tried to insist that Townes stop this nonsense and the waste of government money, for Townes had already then spent about $30 000 under the Joint Services grant administered by the Signal Corps, the US Office of Naval Research and the Air Force.

Finally, one day in April 1953, Jim Gordon rushed into a spectroscopic seminar that Townes was attending crying: it works! The story goes that Townes, Gordon and the other students (Zeiger had by this time left Columbia to go to the Lincoln Laboratory and T C Wang had replaced him) went to a restaurant both to celebrate and to find a Latin or Greek name for the new device, the latter without success. Only a few days later, with the help of some of the students, they coined the acronym MASER: Microwave Amplification by the Stimulated Emission of Radiation. This name appears in the title of a paper in *Physical Review* but detractors interpreted it as the Means of Acquiring Support for Expensive Research!

The first mention of the operation of the oscillator is in a report of 30 January 1954, in nearly the same form in which it was published in a letter to *Physical Review*.

A block diagram of the apparatus is shown in figure 42. A beam of ammonia molecules emerges from an aperture made in a small oven where an accurately controlled temperature is established. The beam contains molecules both in the lower and upper state with a small excess of molecules in the lower state, as is natural, and enters a system of focusing electrodes. These electrodes establish the inhomogeneous field that separates the molecules. They are positioned in such a way that the molecules in the upper state continue to move along the axis of the system, while the molecules in the lower energy state are turned away from the central axis. An accurate design of the separator enables not only the molecules to be separated but

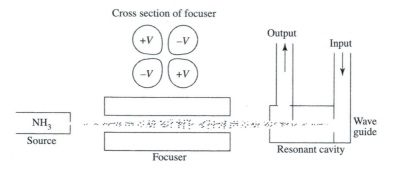

Figure 42. Layout of the ammonia maser.

also those in the upper state to be focused in a well collimated beam. This beam enters the resonant cavity whose resonance frequency is accurately tuned to the transition frequency of ammonia, that is 23 830 MHz. If a sufficient number of molecules in the upper state arrives, a continuous oscillation is obtained that can be extracted from the cavity with the usual microwave techniques; otherwise the system may be maintained in conditions such that there are not enough molecules to oscillate but sufficient to amplify an external signal. In this way the device operates as an amplifier, of course at the same frequency. The entire system is then maintained in an enclosure (not shown in the figure) in which a high vacuum is produced to prevent the ammonia molecules colliding with air molecules and thus returning to the lower state as a result of energy exchange. Of course the whole system is not as simple as it appears from this elementary description.

The principal characteristic of the maser is its extremely low noise, both as an amplifier and an oscillator. This means that the signal is very pure and clean and that all photons are emitted in a coherent way. Only very few photons are emitted randomly by those molecules that de-excite spontaneously and not by stimulated emission and these are the photons which constitute the noise. In many electronic devices noise arises from fluctuations in the number of electrons that produce the electric current. These fluctuations are proportional to temperature and are independent of the specific device; therefore engineers have adopted the habit of characterizing the noise of devices by quoting their noise equivalent temperature, that is the temperature at which an electrical resistance should be brought such that electrons that travel through it produce the observed fluctuations. While for the resistance of an ordinary circuit the noise temperature is practically room temperature (that is 300 K), for the maser the equivalent noise temperature is very low, of the order of only a few degrees Kelvin.

Townes understood immediately that one important application of molecular beam masers would be molecular spectroscopy. Molecular beams had already been considered by gas spectroscopists in the early 1950s; however, the basic problem had been that, as a consequence of beam formation, the resultant density of molecules in the spectrometer cell was extremely low. For beams in which the molecules may be assumed to be in or very near to thermal equilibrium, the absorption and emission processes in the presence of external radiation nearly balance out against a small net absorption which occurs because more molecules are in the lower energy state. In the beam obtained by Townes, with all molecules selected in the required energy state, an enhancement of the signal by up to a factor of one hundred could be obtained. So Gordon used the maser for spectroscopy studies.

The power of the first maser was only 0.01 μW. It was very meagre, but the device emitted a very narrow line. To find how pure the frequency was,

Townes and his group built a second maser to make a comparison between the frequencies emitted by the two masers and early in 1955 they could say that during a time interval of 1 s the frequencies of the two masers differed between them by only 4 parts in 10^9 and over a longer time interval of 1 h by one part in 10^9.

These results suggested the maser as an optimal candidate to make very precise standards of frequency or to build atomic clocks. Research on the ammonia maser became more widespread among other laboratories in universities, government and industry, both under the push of military requests and through personal contacts between interested scientists; however, even so, there were only around ten groups with few researchers and very modest support.

The Russian approach to the maser

Whilst these exciting results were obtained in the USA, what was happening on the other side of the world in the then Soviet Union, full of military secrets and protected by the impenetrable iron curtain? In Moscow, not far from the city centre, one of the largest Physics Institutes of the country existed and still exists today, named after one of the most renowned Russian scientists, the famous Lebedev Physics Institute (shortened by Russians to FIAN), where thousands of researchers worked, administered by the very powerful Russian Academy of Sciences. The Institute had been established in the early 1930s by the physicist Sergej Ivanovic Vavilov (1891–1951). Vavilov is known for his contributions to optics and especially for his studies on luminescence, that is the emission of light by certain bodies upon illumination. Vavilov enounced the laws of the phenomenon. In 1934, when Pavel Alekseevic Cerenkov (1904–1990), a student working on his doctorate thesis, discovered that electrons travelling in water emitted a faint blue radiation, Vavilov, who was his supervisor and had helped him in the measurements, concluded that the phenomenon was not due to luminescence and ascribed it to electrons. Under his push, two researchers of the Institute, Ilja M Frank (1908–1990) and Igor E Tamm (1895–1971) provided the full theoretical interpretation of the phenomenon, showing that charged particles that move in a medium with a speed greater than that of light in that medium, emit radiation. This result led them to the award of the Nobel prize in 1958 together with Cerenkov. The Lebedev Institute has produced in all six Nobel prizewinners and is well known all around the world. Vavilov was its director until his death. As a science organizer and president of the Soviet Academy of Sciences, Vavilov made enormous contributions to the scientific growth of the Soviet Union during the difficult period immediately after the end of the Second World War. At the time of our story, two researchers of the Institute, A M Prokhorov and N G Basov, were interested in solving spectroscopy problems using molecular beams.

Aleksandr Michailovich Prokhorov was born on 11 July 1916 in Atherton, Australia, to the family of a revolutionary worker who emigrated there from exile in Siberia in 1911. Prokhorov's family returned to the Soviet Union in 1923. In 1939 he graduated from Leningrad University and went to work at the Lebedev Institute of Physics, Moscow, beginning his scientific career by studying the propagation of radio waves over the Earth's surface.

During the Second World War he was wounded twice and returned to the Institute in 1944. After the war, following a suggestion by V I Veksler (1907–1966), he demonstrated experimentally in his doctoral thesis that the synchrotron (an elementary particle accelerator used by physicists for high energy experiments) can be used as a source of coherent electromagnetic oscillations in the centimetre waveband. After earning his PhD, he went on to head a group of young research workers (amongst whom was Basov) working on radio wave spectroscopy.

After his work in the field of masers and lasers, which we shall consider in a moment, he was in 1960 elected a Corresponding Member of the USSR Academy of Sciences and in 1966 he became Full Member. For his research he was awarded the title of Hero of Socialist Labour, the Lenin prize and, in 1964, together with Basov and Townes, the Nobel prize. He died January 8 2002.

Nikolaj Gennadievic Basov was born on 14 December 1922 in Usman near Voronezh, about 480 km south of Moscow on the river bearing the same name in central Russia. At the start of the Second World War he graduated from a secondary school in Voronezh, and enlisted. He was sent first to Kuibyshev and then to the Kiev school of Military Medicine, from which he graduated in 1943 with the rank of lieutenant in the medical corps. His service began in the chemical warfare defence forces and then continued at the front. A little after the end of the war, following his return from Germany, he realized his dream of studying physics while still in the Soviet Army. He enrolled at the Moscow Institute of Mechanics (now of Physics Engineering). Exactly 20 years later he was elected to the USSR Academy of Sciences.

In 1948 Basov began work as a laboratory assistant in the Oscillation Laboratory of the Lebedev Institute of Physics. The laboratory was headed at that time by M A Leontovich and Basov later became an engineer there. Later still, in the early 1950s, he entered Prokhorov's group. After the maser work he also made important contributions to the development of a number of lasers. He became director of the Lebedev Institute in 1973 and was also a member of the Supreme Council of the USSR and the Presidium of the Russian Academy of Sciences. He died on 1 July 2001.

At this time Prokhorov was heading research on synchrotron light, that is the light emitted by the electrons or protons accelerated in the circular trajectory of this machine, and Basov started working on this project by studying the emitted radiation. Then Vavilov, who at this time was the director of the Institute, asked them to become involved in microwave spectroscopy. So they built a spectroscope and when it worked started a number of experiments, among which was the study of certain radioactive nuclei.

The group was interested in the molecular spectroscopy of vibrational and rotational states. Everybody was concerned about the sensitivity of the spectroscope being very low. The upper and lower level populations at centimetre wavelengths are nearly the same and the relative difference is of the order of one part per thousand. The group was hoping to obtain population differences capable of giving a sensitivity one thousand times larger. They were also studying the possibility of using microwave absorption spectra to produce frequency and time standards. The accuracy of a microwave frequency standard is determined by the resolving power of the radio wave spectroscope. This in turn depends exclusively on the width of the absorption line itself. An effective way of narrowing down the absorption line was found to be to use spectroscopes operating in conjunction with molecular beams. However, the performance of molecular spectroscopes was strongly limited by the low intensity of the observed lines, which in turn was determined by the small population differences of the transition investigated at microwave frequency. The idea that it was possible to increase the sensitivity of the spectroscope appreciably by artificially varying the population in the levels arose at this stage of their work The use of the separating effect due to the different deflection produced by electric or magnetic inhomogeneous fields was considered. For this purpose a molecule with a large dipole moment was needed and they selected caesium fluoride (CsF). At the same time they understood that for studying the energy levels of the molecules, besides the absorption process traditionally used in spectroscopy, they could also use the emission of excited molecules. The transmission of a beam of molecules in the upper state through a resonator, so that the field generated by the beam reacts with the beam provoking an oscillation, was described by them theoretically at a Soviet Conference on radio spectroscopy (that is an internal conference in which at the time only the Soviets participated) in May 1952, as Prokhorov and Basov affirmed in a review paper written in 1955. At this conference they discussed the possibility of exciting in this way the CsF molecules. In the discussions that followed also ammonia, well known to spectroscopists all around the world, was suggested.

The first published paper by Prokhorov and Basov was sent to the *Journal of Experimental and Theoretical Physics* in Russian in December 1953 and was printed in October 1954, therefore after the publication of Townes' paper on the maser. The delay to the publication occurred due to the desire of its authors to correct some numerical errors in the formulae. It contained a detailed theoretical study of the use of molecular beams in microwave spectroscopy. The authors showed that molecules of the same kind, present in a beam containing molecules in different energy states, can be separated one from the other by letting the beam pass through a non-uniform electric field. Molecules in a pre-selected energy state can then be sent into a microwave resonator where absorption or amplification (according to which energy state has been selected) takes place. Prokhorov and

Basov presented also the quantitative conditions for operation of a micro-wave amplifier or oscillator that they called a 'molecular generator'.

Notwithstanding the isolation—few physicists were allowed to travel abroad and rarely was permission granted for this—the Russian researchers were fully informed of research in the rest of the world through scientific journals. Basov, who for his doctorate thesis worked actively in the new field of quantum radiophysics, as it was called in Russia (in the United States it was called quantum electronics), as soon as he read Townes' letter announcing the construction of the maser, assembled a few months later the first Soviet maser.

Personal contacts were established later. Prokhorov met Townes for the first time in Great Britain in 1955 at a conference of the Faraday Society, where he presented the work we already mentioned, while Basov met Townes, Schawlow, Bloembergen, and many others at the first International Conference on Quantum Electronics at Shawanga Lodge, in New York State, in September 1959.

As we may see, at variance with what occurred in America, neither Basov nor Prokhorov were familiar with radar, nor had worked on it. They arrived at the maser concept from the spectroscopy side and the generic wish to create new kinds of sources in the centimetre wavelength range, which was their primary purpose when working on synchrotron light. In this respect, the Russian school tradition which encourages new ideas without worrying about an immediate practical realization, was very helpful to them.

When the first maser was assembled in Moscow, a stream of visitors from all over the Soviet Union came to see it, and the group built three masers to study their frequency stability. Also a maser was built with two beams that originated from opposite directions which allowed a stability of one part in 10^9 to be obtained. This stability was employed to build a frequency standard which with some improvement was used for a long time at the All Union Institute of Physicotechnical and Radio Engineering Measurements which provided the time standard in the Soviet Union.

The three-level maser

As we will see, the molecular beam maser, notwithstanding some exceptional characteristics, is not very useful for practical applications and probably, if the developments we will describe below had not occurred, the invention by Townes and other researchers would not have received much attention, except in the scientific field of spectroscopy. The principal disadvantage of the ammonia maser, when one wanted to use it for applications other than for frequency standards, was that it emitted a very narrow line (that is at a very well defined frequency) and that the frequency of this line could not be changed to tune the device to other frequencies.

The simple way to increase both the band of frequencies that could be amplified by the device and its ability to change the central frequency of this band (that is its tuning) was to use a different material. An interesting class of transitions was made by the transitions between magnetic levels of ferromagnetic or paramagnetic substances. We have already said that the energy level of an electron in an atom, when the atom is submitted to a magnetic field, splits into many sub-levels: this is the effect discovered by Zeeman. By changing the strength of the external field which produces the magnetic levels, the separation between them changes and therefore frequency tuning becomes possible. Using solids instead of gases, the power is also enormously increased because now the paramagnetic ion concentrations in a solid could easily be one hundred thousand times greater than the number of ammonia molecules in a beam.

Townes took a sabbatical year in 1955–56 that he spent half in Paris and half in Tokyo. When he arrived at the Ecole Normale Supérieure, in the autumn of 1955, one of his ex-students, Arnold Honig, who was working there with Jean Combrisson on electronic paramagnetic resonance, told him that arsenic ions in silicon crystals at liquid helium temperature, had a very long relaxation time of 16 s. Townes immediately realized that such circumstances would allow the arsenic ions to remain in the magnetic state of higher energy long enough to permit energy to be extracted from them by stimulated emission, and persuaded Combrisson and Honig to perform an experiment. When Townes left for Japan in the spring of 1956 the device did not work but Townes was convinced it was a promising direction to explore and the three researchers published a paper in which they discussed the possibilities offered by the system.

At about the same time, but independently from Townes, a physicist at MIT, Malcolm Woodrow P Strandberg (1919–) was considering the possibility of building a maser using solid materials instead of a gas. During the war he had worked with radar and later became interested in microwave spectroscopy, starting in the early 1950s to work on paramagnetic resonance. On 17 May 1956 he ended a seminar at MIT on paramagnetic resonance with some observations on the advantages of a solid state maser. Among the listeners was a young Dutchman, Nicolaas Bloembergen, professor at the Engineering and Applied Physics Department at Harvard.

Bloemergen was born in Dordrecht, the Netherlands, on 11 March 1920. He studied under L S Ornstein (1880–1941) and L Rosenfel (1904–1974) and received the Phil.Cand. and Phil.Drs. from the University of Utrecht in 1941 and 1943, respectively, during the German occupation of the Netherlands. He then escaped to the USA and went to Harvard where he arrived six weeks after Purcell, Torrey and Pound had detected nuclear magnetic resonance. They were busy writing a volume for the MIT Radiation Laboratory series on microwave techniques, and the young Bloembergen was accepted as a graduate assistant and asked to develop the early NMR apparatus, so he started to study nuclear magnetic resonance and in the meantime attended lectures by J Schwinger (1918–1994), J H van Vleck, E C Kemble (1889–1984) and others.

He returned to the Netherlands for a short period after the war and pursued his research within a postdoctoral position at the Kamerling Onnes Laboratory in 1947–48 at the invitation of C J Gorter, who was a visiting professor at Harvard during the summer of 1947. In 1948 he received a PhD at Leyden University with a thesis on nuclear paramagnetic relaxation which was subsequently published as a short book. He then went back to Harvard and joined Purcell and Pound in work on magnetic resonance to which he made important contributions, some of which were referred to in the preceding chapter. In 1951 he became an Associate Professor and in 1957 Gordon McKay Professor of Applied Physics at Harvard, where he has been the Gerhard Gade University Professor since 1980. His important research in the fields of nuclear magnetic resonance, masers and nonlinear optics led to the award of the 1981 Nobel prize for Physics (an award shared with Shawlow and Siegbahn).

After the Strandberg seminar, Bloembergen asked him why he was thinking of a solid system that could not have the spectral purity of the ammonia maser. Strandberg explained he was considering a completely different application for a microwave amplifier with very low noise. Bloembergen was struck by this idea and discussed it with Benjamin Lax, the Head of the Solid State Physics group who introduced him to the work of Combrisson, Honig and Townes. Both in this work and in the idea of Strandberg, a two-level maser was considered. These devices had to be pulsed to operate, and solids with abnormally long relaxation times were required. A device which did not have these limitations would clearly be most useful, and Bloembergen spent some weeks thinking about how it could be realized.

The knowledge Bloembergen had of the behaviour of matter under magnetic fields allowed him to understand that the most suitable levels to be used for the device were not two naturally pre-existing levels in a molecule but the levels that are created when a substance is submitted to a magnetic field (these are the Zeeman levels that we have already considered when speaking of electronic paramagnetic resonance). Therefore Bloembergen considered using the effect of a magnetic field (that is the Zeeman effect) to select at will the two levels between which to produce the transition, tuning the emission frequency by selecting the energy difference between the two levels. In this way he realized that if instead of using two levels he had employed three levels, he needed no longer to separate physically the molecules in the upper energy state, but could select the population of the different levels by a clever use of interactions. To obtain this result he considered atoms hosted in a solid as impurities. Impurity atoms replace some of the atoms that form the solid and find themselves isolated among many foreign atoms. When this happens the orbits of their electrons are perturbed very

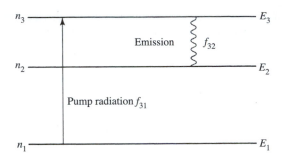

Figure 43. Three-levels configuration of a paramagnetic material.

little with respect to the case in which the atom is in a gaseous state and its energy levels can be considered to be well separated by those of the other atoms of the solid.

To understand Bloembergen's proposal let us recall that atoms or ions with n unpaired electrons (that is with opposite spins) in an external magnetic field acquire $n + 1$ magnetic levels whose splitting is proportional to the strength of the magnetic field (this is the anomalous Zeeman effect that we recounted in chapter 4). Let us now consider a material having three unequally-spaced energy levels (figure 43). Some paramagnetic ions in suitable crystals possess such levels. If we use n_1, n_2 and n_3 for the populations of the three energy levels E_1, E_2, E_3, respectively, at thermal equilibrium we have

$$n_1 > n_2 > n_3.$$

With normal magnetic fields the energy differences among levels are rather small and correspond to microwave frequencies. They are also rather small with respect to the thermal energy of atoms and therefore the three populations are not very different from each other.

The system is now subjected to a strong pumping radiation at a frequency f_{13} which corresponds to the energy difference between level 3 and level 1. Such a field, that we will call the pumping field, is obviously absorbed and induces transitions between the levels 1 and 3. Because, initially, more atoms are in the fundamental level 1, the system will absorb energy populating level 3 at the expenses of level 1. The net effect is that the populations n_1 and n_3 tend to become equal, with n_1 decreasing and n_3 increasing. The population n_2, on the other hand, is not influenced by the field and therefore remains the same. Initially, it was slightly larger than n_3 but eventually, due to the pumping field, n_3 is increased at the expense of n_1 and a situation may be reached in which n_3 is larger also than n_2, or n_2 is larger than n_1. A population inversion between these levels is created and stimulated transition may occur at a frequency f_{32} or f_{21} corresponding to the energy difference between level 3 and 2 or between 2 and 1, respectively. Of course to obtain a sufficiently strong emission, the population differences between the various levels must be as large as possible and for this, because the levels differ from each other by a very small energy, it is necessary to work at a very low temperature.

Bloembergen analysed mathematically the different processes that were occurring and concluded that inversion could be obtained, for example between levels 3 and 2, if the time employed by the atoms in the levels to come back to their fundamental state, the so-called relaxation time, satisfied certain conditions.

At this point we have to say that the idea to use a three-level system came also to Basov and Prokhorov. In 1955 they published a proposal considering gas molecules with three levels and showing the possibility of obtaining inversion using suitable radiation fields. However, the system considered by the two Russian researchers was not tuneable unlike that of Bloemergen. Moreover no discussion was given on the importance of relaxation and none of the proposed methods has ever worked.

Coming back to the USA, at Bell Telephone Laboratories a group with Gordon (who had joined Bell Laboratories after his thesis with Townes) and G Feher, obtained the maser effect in a silicon sample doped according to the scheme suggested by Combrisson, Honig and Townes. A little after, Rudolf Kompfner (1909–1977), the electronics research director at Bell and the inventor of the travelling-wave tube, approached H E Scovil, a member of the technical staff working on the development of solid state devices. The two men had met in Oxford University where Kompfner had worked and Scovil had earned a doctorate in physics studying electronic paramagnetic resonance. Kompfner suggested to Scovil that he should study how to make a solid state maser which operated continuously. On 7 August 1956, Scovil presented a memorandum with the proposal to employ a gadolinium ethyl sulphate crystal, whose properties he had studied in detail at the time of his thesis, using its paramagnetic levels according to an approach identical to that proposed by Bloembergen, and prepared a paper to be sent to the *Physical Review*.

Notification of Bloembergen's work had started to arrive at Bell Laboratories, and Bloembergen had heard that something was happening at Bell. Bloembergen was willing to patent his maser and started to worry that he had released too much information to colleagues. On the other side at Bell Laboratories, it was feared that an unpleasant situation would arise with claims on the origin of ideas and future litigation for violation of patent. So Bloembergen was invited to present his results at Bell Laboratories and on 7 September 1956 he gave a seminar in the New Jersey centre. Scovil, who did not know about Bloembergen's work when he attended the seminar, understood that, in his own words, 'Bloembergen had the same idea and effectively had it before me. So I did not send my work for publication.'

Instead Bell Laboratories negotiated an agreement to have the use of Bloembergen's patent, leaving the two parties free to compete amicably to realize the first experimental maser of this kind.

In the meantime Bloembergen had published his proposal in *Physical Review* in a paper received on 6 July 1956 which appeared in the 15 October issue of the same year. In it he considered additionally some possible materials which could be employed for making a maser.

Unfortunately he and his group at Harvard were interested, for astronomical purposes, in a device working on the interstellar hydrogen line at 1420 MHz and therefore chose a material which could work at that frequency, failing to obtain the first successful operation of the three-level maser. The following year after publication of Bloembergen's theoretical work, the first three-level maser was built (1957) at Bell Telephone Laboratories by H E Scovil, G Feher and H Seidel by utilizing gadolinium ions in a lanthanum ethyl sulphate crystal. Shortly after (1958), Alan L McWhorter and James W Meyer at MIT Lincoln Laboratory used chromium ions in a cobalt and potassium cyanide to build the first amplifier. Bloembergen and collaborators eventually succeeded in making their maser work, but came third in 1958.

In building their maser, Scovil and collaborators used a clever operating principle. How much gadolinium ions amplify depends on how much population inversion is obtained between the maser levels. In the gadolinium case the levels were 2 and 1. The population difference between the two levels depends, among other things, on how fast the ions pumped from level 1 into level 3 decay into level 2. The group observed that, in their crystal, caesium was present as an impurity which by interacting with gadolinium increased the velocity with which the ion decays from level 3 to level 2. Accordingly they chose a concentration of caesium atoms so as to optimize the energy transfer between these two levels.

While the original ammonia maser was principally useful as a frequency standard, due to the stability of its emission frequency, or else as a very sensitive detector, the solid state maser, being tuneable, was something which really could be used for communications and radar. It could be tuned in a continuous way on a reasonable band of frequencies, still maintaining the principal low noise characteristic of the maser. Tuning could be obtained by changing the magnetic field strength.

Not much later, Chihiro Kikuchi and his co-workers showed that ruby was a good material for a maser. In 1955 the engineer Weston E Vivian had started research at the Willow Run Laboratories at Michigan University, supported by the military, for a passive detection system in which, with a very sensitive detector, the microwaves naturally emitted by objects were detected (remember the blackbody law of chapter 3). Vivian calculated that microwave receivers of extraordinary sensitivity were needed. Kikuchi, initially engaged in studying the microwave absorption properties of crystals, was asked to try to build a good maser which served the purpose and, after having considered chromium trichelate, which technicians had difficulty growing, thought about using pink ruby.

Ruby is a colourless crystal of an aluminium oxide (Al_2O_3) in which chromium atoms are introduced as dopants. These atoms replace some of the aluminium atoms, lose three of their valence electrons, acquiring therefore three positive charges and, as we will see later, are responsible with their optical behaviour for the beautiful red colour of ruby. Of course, the rubies used for maser construction are synthetic. The intensity of the coloration depends on the concentration of the chromium ions. The higher this is, the more intense the red coloration of the ruby.

In January 1957, Kikuchi, having obtained a sample of pink ruby from the Mineralogical Department of the University, started to design the maser.

An important parameter in the construction of a maser is the angle that the external magnetic field makes with the crystal axis. The energy separation between the levels, and therefore the operation frequency of the maser, depend on it. At the time, the preferred angle was about 15°. But at that angle, to calculate the position of the energy levels required the use of a computer, which was not as easy to find as it is today. Kikuchi chose instead a different angle of 54° 44' (figure 44). For this angle the equations simplified so that analytical solutions could be obtained that showed that a maser could be built at the wavelength of 3.2 cm, which was familiar to the laboratory technicians through their radar work.

However, the work progressed slowly, and only on 20 December 1957 did the maser work. After that McWhorter and Meyer at Lincoln Laboratory in the spring achieved a working maser using potassium–cobalt cyanide doped with chromium, Townes and his co-workers had operated a maser at 3 cm, and Bloembergen, J O Artman and S Shapiro with the same material had assembled their own maser at 21 cm.

Potassium–cobalt cyanide is a very poisonous material. In 1958, Bloembergen, Charles Townes and their wives were dining together in New York, and Mrs Townes talked to Bloembergen's wife and showed off a very nice pendant with a ruby crystal set in gold on a gold chain. Her husband had had it made for her in commemoration of the maser. That night in their hotel Mrs Bloembergen asked her husband 'When are you going to give me something related to your maser?' So Bloembergen said 'Well, dear, my maser works with cyanide' and so was saved of a costly gift!

The ruby maser made all the other crystals obsolete. The ruby was an easily obtainable material. It was strong and easy to handle for frequency tuning.

It was Joseph Geusic who became active in the design and perfection of the ruby maser. He had just gone to Bell Laboratories from Ohio State University where he had written his thesis under J G Daunt, dealing for the first time with the measurement of microwave resonance in ruby.

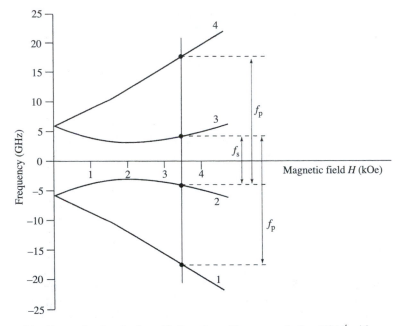

Figure 44. Energy levels of ruby with its axis making an angle $\theta = 54° 44'$ with respect to the magnetic field. (From Thorp J S 1967 *Masers and Lasers: Physics and Design* (London: Macmillan) p 63.)

During 1957 and 1958 many masers were built in several laboratories, including Harvard, using chromium ions in ruby crystals. Rubies were employed in a great number of types of maser with many different characteristics. Since 1958 many masers were built for applications in radio astronomy or else as components in radar receivers. These masers were almost all of the ruby type.

As soon as they appeared, masers attracted great interest from the military who thought of using them as very sensitive radar receivers with very low noise. Also the growing field of radio astronomy considered using them because of their extremely low noise which allowed the detection of very faint signals. However, great inconveniences existed. The maser *per se*, the three-level version which had the more suitable characteristics for these applications, was small and robust, but required cooling to liquid helium temperature and placing in a powerful magnetic field, and the cooling system and the magnet were very cumbersome and heavy (figure 45). The device therefore was inappropriate for the battlefield or for installation within an aircraft. Also for the radio astronomical application, its weight and encumbrance were unfavourable, considering that to fully exploit its low noise characteristics it was necessary to mount it at the centre of a gigantic receiving antenna, because otherwise the components that would have to

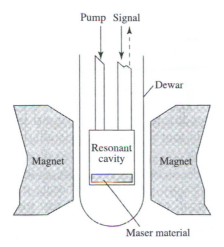

Figure 45. Essential elements of a solid-state maser.

be employed to transfer the signal to a more suitable place would introduce their own noise and effectively cancel the advantages of the device (figure 46).

Also in satellite communication systems, in which satellites were used to reflect back to Earth signals sent from transmitting stations or to re-transmit these signals, masers eventually failed to find great application. The reason was that in the meantime new semiconductor devices had been developed, the so-called parametric oscillators, which even though they did not yield as little noise as masers, were, however, much lighter and manageable and moreover did not require cooling or strong magnetic fields.

Ultimately, maser applications were restricted to a very limited number of cases; however, the hectic activity that developed around them, and the new knowledge they brought, showing for the first time the practical application of stimulated emission, favoured the development of the laser with all the applications that followed.

Masers had, however, a moment of glory; a ruby maser was employed by Arno A Penzias (1933–) and R W Wilson (1936–) to discover in 1965 the blackbody radiation at 3 K from the Big Bang, which earned the two discoverers the physics Nobel prize in 1978 together with P L Kapitza (1894–1984) (a Russian physicist who received it for his studies on low temperatures, discovering unusual properties of helium gas when it was liquefied and inventing a technique to liquefy it that is used still today).

The story is interesting, showing a Nobel prize may be earned almost by chance.

Penzias was born in Munich in 1933 and at the age of six risked being deported with his family to Poland; so the family emigrated first to England and then reached the United States in 1940. Here he earned a degree in chemical engineering and after marriage and two years in the US Army

199

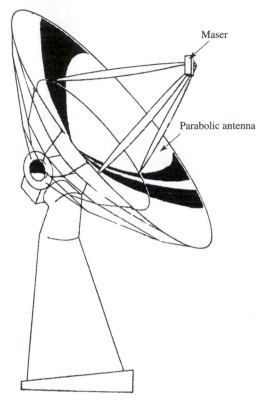

Maser

Parabolic antenna

Figure 46. Typical parabolic antenna used for radio astronomy, telemetry, radar etc. with a maser mounted in the focal point of the parabola.

Signal Corps, he enrolled in Columbia University in 1956 to study physics with I I Rabi, P Kusch and C H Townes. As a thesis, Townes assigned him the task of building a maser amplifier in a radio astronomy experiment of his own choice.

In 1961, after his thesis work, Penzias sought work at Bell Laboratories and considering the exceptional instrumentation there decided the best thing to do was to complete the observations he had started with his thesis work. 'Why do you not accept a fixed position?' Rudi Kompfner, who was then Director of the radio laboratory, asked him. 'In any case you may leave when you wish'. Thus he was recruited and remained at Bell Laboratories until his retirement in 1998.

The project was to detect the still missing emission of the interstellar OH molecule. In the meantime at MIT scientists succeeded in detecting the OH emission, and Penzias moved with his equipment to Harvard to make observations. In mid 1962 the Bell System launched the TELSTAR satellite, and being afraid that the European researchers would not finish their receiving station in due time, prepared at Holmdel a station equipped with a new maser with ultra-low noise operating at the wavelength of 7.35 cm. In the end, this equipment was not needed because the Europeans finished their station on time and Penzias with Robert Wilson, a radio-astronomer from CalTech, could use it.

> Robert Wilson was born in 1936 at Houston, Texas, where his father worked in oil-wells and since childhood was interested in electronics. He took his degree in physics at Rice University. Then he passed to CalTech for a PhD in physics, and became interested in radio-astronomy and after completing his thesis at the Owens Valley Radio Observatory went to Bell Laboratories at Crawford Hill in 1963 where he started work with Arno Penzias in a long and fruitful collaboration.

While mounting the receiving system for radio astronomical observations, Penzias and Wilson started a series of astronomical observations aimed at optimizing the performance of the antenna and the maser, and to do this they measured the intensity of the radiation emitted by our galaxy. They performed very accurate calibration measurements. In 1963 the maser at the wavelength of 7.35 cm was installed and they performed a series of operations to calibrate the whole system. Everything was under control except for the fact that the noise in the full system was 3.5 K larger than the value they had calculated. Penzias and Wilson began an accurate examination of the possible reasons for this difference and eventually, after having eliminated all the other alternative hypotheses, concluded that the extra 3.5 K noise was effectively due to radiation arriving at the antenna uniformly from all space.

One day Penzias spoke of the noise problem to Bernard Burke of MIT who remembered a theoretical study on the radiation in the universe by P J E Peebles from the group of Professor R H Dicke in Princeton. Penzias called Dicke who sent him a copy of Peebles' work. In the paper, Peebles following Dicke's suggestion calculated that the universe should be filled with a relic blackbody radiation at a minimum temperature of about 10 K, reminiscent of the primeval explosion (the Big Bang). In 1948 George Gamow had already performed calculations on the initial conditions of universe. The Big Bang model assumes that the universe was born in a gigantic explosion. Immediately after, the temperature should have been extremely high, of the order of 10 thousand million degrees, or maybe more. At such a temperature, of course, matter does not exist but only a broth of protons, neutrons, electrons, photons and other elementary particles. These particles by interacting with each other started to produce the lighter elements while a great quantity of very energetic radiation was emitted and the expanding universe started to cool. At an age of less than a few hundred thousand years, the matter in the universe was still ionized and interacted strongly with light. At this time the temperature was decreased to about 3000 K and the electric charges of matter started to recombine, making matter neutral. At this point the interaction of elementary particles with photons ceased and from that time the electromagnetic radiation of the universe started to cool due to the expansion of the universe with no further interaction with elementary particles, shifting the wavelengths towards higher values while the number of photons per unit volume (that is their density) decreased. Due to one property of the expansion, the Planck distribution was related to a temperature that decreased proportionally with the Universe's dimension. Small variations in the intensity trace small perturbations in the density of primordial matter, which have been amplified by gravitational forces to form the galaxies.

At the time of our story, the problem of this radiation was being discussed again by astrophysicists after having been forgotten for some time, and the Dicke group was very interested. After the first contact, Dicke and his co-workers visited Penzias and Wilson and convinced themselves of the reality of the measurements. Therefore two letters were sent and published in the *Astrophysical Journal*, one signed by Penzias and Wilson announcing the discovery, and another one signed by Dicke, Peebles, P G Roll and D T Wilkinson in which a theoretical explanation was given.

For this discovery Penzias and Wilson were awarded the physics Nobel prize in 1978. It is largely needless to say that their discovery was made possible by the use of the maser that has such a low noise that it is sensitive

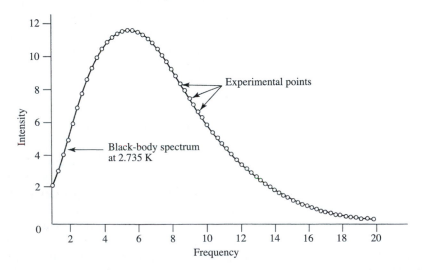

Figure 47. Spectrum of the cosmic microwave background radiation measured with the COBE (Cosmic Background Explorer) in 1989. Points represents experimental data. The continuous curve is the spectrum of a 2.735 K theoretical Planck blackbody.

to the very low noise of the radiation that they measured. More precise measurements today yield 2.735 K for the value of the temperature of this radiation and its existence is part of the experimental evidence for the Big Bang model (figure 47). But why just 2.735 K and not another value is one of the most important problems of modern cosmology, pertinent to some of the fundamental aspects of the structure and evolution of the Universe, and which is awaiting an answer.

Atomic clocks

One of the more interesting applications of atomic beam masers was found in the construction of atomic clocks. Very precise clocks can be used to establish if the astronomical 'constants' stay fixed or change with time and may allow the general relativity theory to be verified. Besides their scientific meaning, atomic clocks have important military and commercial uses. In the 1950s and 1960s precise frequency standards were required for earth-based stations and in planes using long-distance aerial radio-guided navigation system that at that time was in the final stage of development by the US Army Signal Corps. Highly stable frequency standards resistant to vibration were also part of the electronic equipment mounted on teleguided missiles, research which was financed by the military.

The maser is an optimum frequency standard that promised better precision with respect to the atomic clocks already created. For this purpose the hydrogen maser was very useful. It was developed by Ramsey with Daniel Kleppner and Mark Goldenberg (from MIT) in 1961 and was the first atomic maser. Its very precise emitted frequency was used to stabilize the microwave oscillator in Ramsey's two-field system.

The hydrogen maser (which oscillates at a frequency of 1420 MHz) was used in 1976 to verify the predictions of general relativity with an accuracy of one part in 10 000 by comparing the frequency of a hydrogen maser clock at rest on Earth with that of another similar clock travelling in a missile. Relativity theory states that absolute simultaneity does not exist. If a stationary observer measures the time between two events and finds it to be for example 60 s, another observer in motion with respect to the first finds that this time is longer. The clock in motion is delayed with respect to that of the Earth-bound observer.

Hydrogen masers accurately synchronized were also used to guide the Voyager II in its historic encounter with Neptune.

Oscillations from accelerated electrons

Early in 1951, the physicist Hans Motz (1909–1987) proposed a new way of producing radiation at millimetre and sub-millimetre wavelengths that did not make explicit mention of the population inversion processes or of

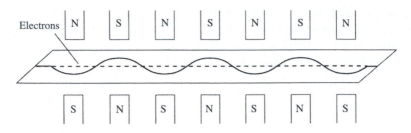

Figure 48. Free-electron maser or laser. The electron beam travels across a series of magnets whose fields are alternating in orientation (N and S stands for North and South pole). As a consequence electrons move in circle arcs lying in the plane orthogonal to the field and radiate electromagnetic waves (the radiation is not shown in the figure).

stimulated emission, even if these concepts are implicit in its operation. Later the device evolved into one of the many ways of producing laser radiation with the name of the free electron laser, and is today one of the few lasers emitting at very short wavelengths.

Motz's proposal was made in 1951 when he was at the Microwave Laboratory of Stanford University, near San Francisco, California. His idea was to make a beam of electrons travel through an array of magnets with alternating polarity, that is North, South, North etc. Under the action of the magnetic field the electrons no longer travel along a straight trajectory but along an arc of a circle. After a while the beam finds itself under a field directed in the opposite direction because the polarity of the magnets has been inverted, so the arc bends in the opposite direction and the electron beam travels through a series of semicircular trajectories as shown in figure 48. The electrons that travel on these curved trajectories, according to the laws of electromagnetism, that we already mentioned speaking of the planetary model of the atoms proposed by Rutherford, emit radiation. Under suitable conditions the radiation emitted by the different segments may generate a continuous train of waves. Because the electrons in the beam travel at very high speed, relativity theory shows that through a double interplay of relativistic contractions, the wavelength of the emitted radiation is linked to the radius of the semicircular orbits made by the electrons but is much shorter, falling in the millimetre or sub-millimetre range or, by a suitable design even in the visible and shorter wavelengths.

The interesting point of such a device is that by changing the energy of electrons and therefore their velocity or the distance between the alternate poles of the magnets, it is possible to change the wavelength, so obtaining a tuneable source.

Motz gave an experimental demonstration in 1953, still at Stanford, using a linear accelerator and obtaining several watts of power at the wavelength of about 1.9 mm. It does not seem that there was ever any interaction with Townes.

Celestial masers

The reader at this point could have thought the maser to be an invention of mankind. Nature, however, as often happens, shows that there is nothing new under the Sun! Some years ago in radio astronomy, scientists started to make observations at the frequency of 1420 MHz, that corresponds to the emission of gaseous hydrogen in the interstellar clouds. This particular radio emission represents the spontaneous emission of a particular transition in atomic hydrogen coming from hydrogen atoms at thermal equilibrium at a fairly low temperature (less than 100 K). As such it has none of the characteristics of maser amplification. The study was also extended to other frequencies and led to the identification of the presence of various interstellar gases.

In 1965 the Radio Astronomy Group led by professor H Weaver at Berkeley, California, observed a radio emission at about 1670 MHz coming from OH molecules located near some stars. This emission, in reality, consisted of four known OH transitions at 1612, 1665, 1667 and 1720 MHz. If the emission of these lines had originated spontaneously, the four lines would have had intensities in the ratios $1:5:9:1$, as predicted by the well known transition intensities for the four transitions. However, the observed intensity ratios were quite different and changed fairly quickly with time (within a time scale of months). The distribution of the emitted frequencies of these lines was not a smooth one, but sometimes contained very narrow spectral components. The linewidths were such that the temperature of the source would have to be lower than 50 K to have a broadening as small as that observed. At the same time the emission intensity was so strong as to correspond to a source temperature of 10^{12} K, and the emission apparently originated either from extremely narrow point sources or else appeared in the form of highly directional beams.

The only reasonable explanation of these results was that such radiation originated by spontaneous emission in some part of an OH cloud, and was then strongly and selectively amplified by a maser amplifier while traversing other regions of the OH cloud. Such amplification could explain the anomalous intensity ratios and the high intensity and directionality of the emission.

It also seems reasonable that the maser gain properties can change quickly with time, although changes of both the total quantity of OH and of the associated spontaneous emission should not be expected on the time scales observed. The pumping mechanism responsible for the population inversion is now understood. Molecules are excited by the infrared radiation emitted by cosmic dust and, in suitable conditions, a population inversion is created.

In 1968 other substances were found to emit similarly, and today more than one thousand masers have been discovered in our galaxy that utilize more than 36 molecules and nearly 200 transitions. Among the molecules for which emission has been discovered, besides OH, are water, methanol, ammonia and SiO.

Today it is thought that these celestial masers exist in regions where stars are forming or where stars are very near to the end of their life cycle. Both types of stars have in common the presence of strong fluxes of material sent into space. Typical jets have a velocity of about 30 km/s and the most energetic reach 300 km/s. The matter launched into space condenses rapidly and may be pumped by the infrared radiation emitted by the same star.

The emission mechanisms of the various molecules are in some cases the pumping of the jets with infrared radiation, as stated, but in other cases it could be an excitation by collisions. For example, in the case of SiO, results have been found that confirm this idea. Most of the SiO masers are in the outer atmosphere of giant and super-giant, highly evolved stars. Stars of this type lose a great quantity of their atmosphere as wind, enriching the interstellar matter of the galaxy. During this strong wind, molecules of SiO may be excited by collisions with other molecules that find themselves in the high velocity flux of matter that constitutes the wind.

More recently, in the nuclei of more than 50 galaxies, masers have been discovered that are a million times more luminous than the galactic ones. These megamasers, as they are called, seem to function in some cases via an infrared pumping mechanism while in other cases the mechanism is still under discussion.

The study of these masers, besides being interesting in itself, promises to be helpful in the understanding of the astrophysical evolution processes of stars.

CHAPTER 11

THE PROPOSAL FOR AN 'OPTICAL MASER'

Townes in his research set out to build an ammonia maser at a wavelength of about 0.5 mm and then converted to the much longer wavelength of 1.25 cm for reasons of convenience in construction. All the other masers built afterwards operated in the centimetre range. There was no generator based on stimulated emission able to emit in the millimetre and sub-millimetre region. Although other kinds of traditional generators had been improved, like klystrons, magnetrons and the travelling wave tubes that successfully went down to somewhat more than 1 mm, this wavelength was the shortest thought possible for these devices and their power was very low. Indeed there was no real demand for a coherent emitter in the millimetre and infrared region. One of the most important applications—night vision—employs infrared rays emitted by the thermal scene and needs only a good detector but does not need any emitter. For another important application, spectroscopy, ordinary infrared lamps had sufficient power already. Therefore there was no pressure to develop new sources. However, researchers are curious and like to extend the frontiers of knowledge. So even if without any haste, once the maser was built, one started to think towards the application of its working principle to build a light generator then referred to as an optical maser, and the task was taken up independently both in the United States and in the former Soviet Union.

To make an optical device of this kind, energy levels different from those used for the microwave maser have to be considered. The microwave frequency is so low that the energy gaps needed for the emission have to be found in the roto-vibrational states of the molecule or in the fine structures of atoms submitted to magnetic fields, about which we have already spoken. In the optical case the emitted photon must have energy at least one hundred times larger, and therefore must be produced via transitions between the electronic levels of the atoms.

The other essential element of the device is the resonant cavity that is necessary for an efficient interaction both at microwaves and shorter wavelengths. Microwave resonant cavities have dimensions comparable with the wavelength, that is of the order of a centimetre. With accurate machining, very good microwave cavities can be built. In the case of light, the wavelength is of the order of 1 μm (that is one tenth thousandth of a centimetre) or even less; therefore at the time it was absolutely impossible to consider making a resonant cavity of such dimensions, and the problem as usual seemed insurmountable. Without the resonant cavity a sufficient interaction between the radiation and the particles that emit it cannot be obtained, stimulated emission is poor, and the principal properties of the device are lost. However, alternative methods promoting sufficient interaction between the excited particles and the radiation were proposed and a system composed of two semitransparent plane mirrors parallel to each other was considered. This system had been used already in spectroscopy to perform measurements of wavelength with great accuracy. As we will see later, it is a true resonant cavity, although of a different kind from the ones used at microwave frequencies, and it was introduced in 1899 by two French researchers, C Fabry (1867–1945) and A Pérot (1863–1925). Today it is called a Fabry–Pérot interferometer or simply a Fabry–Pérot. In it the radiation travels back and forth between the two mirrors being subjected to interference phenomena so that inside the cavity only certain wavelengths can exist. If radiation containing several frequencies comes into the device, usually made with circular mirrors, by adjusting the position of the mirrors very carefully, one obtains a series of rings whose radii depend on the wavelength. Used in this way the interferometer was employed by Fabry and Pérot to study the frequency distribution of spectra formed by very near frequencies (the so called fine and hyperfine structures).

On the other hand, if an inverted medium is put inside a Fabry–Pérot, the possibility of having in its interior radiation travelling back and forth between the mirrors is just what is needed to increase the interaction between the generated radiation and the excited atoms that emit it, realizing a kind of resonant cavity at optical wavelengths. Although the dimensions of the cavity are now much greater than the wavelength, a good selection of the possible modes in which the radiation may oscillate is achieved because only radiation that travels exactly parallel to the axis of the device is multiply reflected, while that which forms even a small angle between its direction and the plane of the mirrors is immediately expelled after one or two reflections.

The Fabrikant proposal

As we have already observed, at the forefront among the proponents of the idea to build an optical generator that exploited the principle of stimulated emission was Valentin Aleksandrovic Fabrikant (1907–1991) who in the Soviet Union made his proposal in the 1940s.

Fabrikant began his scientific career as a student at the Physics and Mathematics Faculty of the University of Moscow in 1929 with G S Landsberg. After taking his degree, he was employed in the Soviet Institute of Electronics where soon he became head of a laboratory. In 1932 his attention was focused on problems of optics and on the properties of electric discharges in gases. He published a series of papers in which he studied the spectral composition and intensity of radiation emitted during an electric discharge in a gas, investigating particularly the collision processes between the excited atoms of the discharge and the electrons and the energy transfers that occurred in these processes.

An atom or a very fast electron may strike another atom and transfer part of its energy which, if sufficient, is able to excite the target atom, raising it to an excited level. This is referred to as a collision of the first kind. There also exists a different kind of collision, called a collision of the second kind, in which an atom which is already in an excited state collides with another atom which is in the ground state and gives it its energy. The result is that the first atom returns to the fundamental state, and the second atom jumps to an excited level. The two atoms need not necessarily be of the same kind; it suffices that the two excited levels have the same energy. If the two atoms are of different kinds, through this mechanism it is possible that atoms of one type, say A, excite atoms of a different kind, say B, to an excited level in larger number than that which would occur through thermal collisions of the first kind. In this way, B atoms may show a distribution within the different energy levels that departs from the Maxwell–Boltzmann distribution and this possibility was exactly what eventually interested Fabrikant.

In 1939 he started to examine the possibility of obtaining populations of excited atoms larger in number than predicted by the Boltzmann distribution law and came to show that when radiation travels in a medium in which a population inversion has been realized, one should observe amplification rather than absorption. Then he proposed a way of realizing experimentally an inverted population in a discharge of a mixture of gases using collisions amongst the discharge atoms. These results were included in his doctorate thesis that he defended in 1939 and published in 1940. At this time Fabrikant's interest was associated with obtaining an experimental proof of the existence of stimulated emission. Later he considered the problem more extensively and on 18 June 1951, together with some of his students, he submitted a request for a patent on a new method of amplification of light under the title: 'A method for the amplification of electromagnetic radiation (ultraviolet, visible, infrared and radio waves) characterized by the fact that the amplified radiation is made to pass through a medium that by means of auxiliary radiation or by other means generates an excess concentration with respect to the equilibrium concentration of atoms, other particles, or systems at upper energy levels corresponding to excited states'.

In the patent the ideas on the use of stimulated emission for amplifying radiation were further developed and exposed in a more concrete way. It was shown that the passage of radiation in a medium with inverted population bore an exponential increase of intensity and the proposal employed optical pumping in a three-level system and the use of a pulsed discharge, besides the use of collisions in a discharge. However, the patent was only published in 1959 and since in the Soviet Union modifications to the content of the request were allowed during the procedure to obtain it, it is not possible to know what was added between the first application and the date of publication. The title is so general as to cover practically anything connected with maser or laser action. The Russian patent office was not willing initially to accept the request. Although the purpose of the patent appears very general, Fabrikant was principally interested in light emission in a gaseous medium, recognizing the use of enclosing the amplifying material in a resonant structure. After the application for the patent in 1951, he and his students continued experimental work with different materials but without success. Even though they published experimental confirmation of their ideas, this was later disproved.

Although Fabrikant's work is historically interesting because he developed the concept of the laser, arriving at it from the optical side without passing through the maser phase, it had no influence on the developments of either maser or laser because the work was known only after both devices had been realized. Even in Russia his colleagues probably did not take him very seriously and therefore he did not stimulate anybody even there. Only after the first masers and lasers were built was the Government reminded of him. In 1965 the Soviet Academy of Sciences awarded him with the S I Vavilov gold medal for 'the important work on the optics of gas discharges' and he received the prize of the Soviet Socialist Republics for his contribution to the development of luminescent lamps.

The optical bomb

In America, Robert H Dicke (1916–1997) published in 1953 a paper in which he introduced a new concept that he called 'superradiance', which was at the centre of his discussion of a method to obtain visible radiation with characteristics of coherence. Because in 1953 the names maser and laser had not yet been coined, Dicke called his device an 'optical bomb' predicting that it would emit an exceptionally short and intense light pulse.

> Dicke was a physics professor at Princeton University and a consultant at the RCA laboratories at Princeton. He was interested in basic physical problems and made important contributions to the study of gravity, but amused himself as well by inventing things. His consultant position at RCA made him feel obliged to translate some of his ideas into patents or projects for the company laboratory. He was interested in methods of narrowing the emission line of the hydrogen atom to study the interaction of the electron with the nucleus and this study led him to examine more in general the properties of coherent radiation and to introduce the concept of superradiance. Narrow lines were important to build frequency standards and RCA had a

contract with the Army Signal Corps based on Dicke's idea of coherent emission. Later in 1960 he became interested in gravitation and in 1964 in the Big Bang theory, encouraging his colleague Peebles to calculate the blackbody radiation temperature reminiscent of the great explosion (the Big Bang).

Early in February 1956, Dicke in his notebook wrote about three inventions. One concerned the exploitation of ammonia transitions for producing emission in the millimetre region. The second was a method to obtain more power, shaping the source of the ammonia molecules into a circle around the microwave cavity. The third was a proposal to use this 'circular maser' to generate waves in the region from 0.25 to 0.03 mm, that is in the infrared region. To make his maser work in this part of the spectrum, Dicke substituted the microwave cavity with a couple of parallel mirrors, that is with a Fabry–Pérot. For this idea the RCA lawyers, in May 1956, asked for a patent that was obtained in 1958 under the title 'Molecular Amplification and Generation Systems and Methods'. However, the idea was not further investigated.

The Townes and Shawlow proposal

In the meantime in the Soviet Union, Basov and Prokhorov at the Lebedev Institute studied how to extend to the visible region the properties of the maser, and Gordon Gould (1920–) worked in America on a private project of his own, of which we will speak later.

However, Charles H Townes and Arthur Shawlow were the first to publish a detailed and exhaustive proposal which subsequently led to the construction of various types of lasers.

In 1957 Charles Townes began considering the problems connected with making maser-type devices work at optical wavelengths. Townes carried out his work in close collaboration with Arthur Shawlow, then a research physicist at Bell Laboratories.

Arthur L Shawlow was born in Mount Vernon, New York, on 5 May 1921. At first he intended to become an engineer, today we would say an electronic engineer, but when he finished secondary school he was only 16. In Canada, where he was living at the time, only boys over the age of 17 were admitted to University. Moreover his family could not afford to support him at University. So Arthur applied successfully for a fellowship, obtaining one for mathematics and physics, and after enrolling in that Faculty decided to continue studying physics at Toronto University, taking his degree in 1941. He was fascinated by research and after some hesitation decided to take a PhD, which he obtained at the same University in 1949. Some months before, in Ottawa, a Congress of the Canadian Association of Physicists had been held, in which I I Rabi participated describing the wonderful discoveries that Willis Lamb and Polykarp Kusch had made and for which they had been awarded the Nobel prize. The young Shawlow remained fascinated and tried everything to go to Columbia University. So he wrote to Rabi at Columbia University, who suggested that he applied for a post-doctoral fellowship to work under Townes. This fellowship, as we have already said, was given by Carbide and Carbon Chemical Corporation, a division of Union Carbide, for research into the applications of microwave spectroscopy to organic chemistry. Shawlow had no interest in organic chemistry and had never followed an organic chemistry course at University but he was interested in microwaves. As a student he had a klystron in the laboratory and for one year, during the war, he had worked on microwaves. The fellowship involved working with a man named Charles Townes, whom he had never heard of, but he wanted to go to Columbia and so applied for the fellowship and was successful. At this time, no less than eight future Nobel prizewinners were at Columbia: the nuclear physicist Hideki Yukawa (1907–1981) who was awarded with the Nobel a few months after Shawlow had arrived, for the prediction of the existence of the meson, an elementary particle necessary to explain the interactions between particles that constitute the atomic nuclei; Bohr's son Aage (1922–) Nobel prize in 1975 with B R Mottelson and L J Rainwater for their study of atomic nuclei; Townes, Kusch, Lamb, James Rainwater (1917–1986) and Val Fitch (1923–) Nobel prize in 1980 with J W Cronin for their studies on elementary particles. Shawlow was very excited and earned the fellowship in the years 1949–50. The following year he remained at Columbia as a Research Associate. Shawlow and Townes became friends, often dining together at the Columbia Faculty Club where there was a table reserved for a group of physics and mathematics professors. During this period Shawlow met and married Townes' younger sister. After the marriage he found himself in search of a job. His wife wanted to remain near New York where she was learning to sing. Therefore in 1951 Shawlow accepted a post working in solid-state physics at Bell Laboratories. John Bardeen (1908–1991), one of the inventors of the transistor Nobel prize twice, once in 1956 with W B Shockley (1910–1989) and W H Brattain (1902–1987) 'for their research on semiconductors and their discovery of transistor effect', and later in 1972 with Leon N Cooper (1930) and R Schrieffer (1931) for 'their jointly developed theory of superconductivity', had become interested in superconductivity and wanted someone to help him in his experiments. Shawlow, at this time, had no experience of low temperatures or of superconductors but accepted in any case. To cap it all, Bardeen, when Shawlow arrived at Bell, had decided to move to Illinois University and so Shawlow found himself working in a field he knew little about. At this time he was also writing a book with Townes on microwave spectroscopy and he spent nearly every Saturday at Columbia. After the invention of the laser Shawlow was persuaded to move to Stanford University as a Physics professor, because the oldest of his three children was autistic and he felt there were better institutions for autism in California. At Stanford he developed many new spectroscopic techniques using lasers, particularly for high spectral revolution and high precision. Together with Theodor W Hansch, who joined his lab in 1970, he developed new techniques in laser spectroscopy and the proposal for laser cooling of atomic gases. These contributions were cited when Shawlow shared the 1981 Nobel prize in physics with Nicholas Bloembergen and K M Siegbahn for contributions to laser spectroscopy. He had a good sense of humour and comedic skills. One of the sketches he often liked to present in public representation was to pull out his famous red toy gun in which a small ruby laser was installed. Next, he noisily began to inflate a large clear balloon, inside which a blue balloon with big ears like Mickey Mouse was also inflating. 'There's a mouse inside the balloon' Shawlow was saying. 'You know it is terrible the way mice get into everything. We have to use our laser.' And he flashed the ray gun that burned the inner mouse-shaped balloon while the other balloon remained unharmed. He then explained 'Now this is a very serious experiment. It illustrates how, with lasers, light is no longer something to look with, it is something you can do things with, and you can do them in places where you can see, but not touch.' Shawlow died on 28 April 1999 from congestive heart failure brought on by leukaemia.

In September 1957 Townes had written a draft with the general idea of what could be the first maser operating at optical frequencies, the so-called optical maser. It consisted of a system in which the active medium was illuminated with radiation of a suitable wavelength (this method has been called optical pumping) placed in a cavity having the shape of a box of about 1 cm in dimension, and therefore much larger than the wavelength of light which is about half a millionth of a metre ($0.5\,\mu m$), with part of the walls removed so that the light could enter for excitation purposes and the others silvered so as to reflect the emitted radiation. He had already seen some defects of such a device, but was hoping it could work. The active substance could be a gas and Townes initially had thought of thallium vapour, which could have been excited by illuminating with a suitable light source.

On the 14th of that month, Townes asked a Columbia graduate student, Joseph Anthony Giordmaine, to sign a notebook wherein a light resonator was described: this consisted of a glass box with four mirrored walls, and used a thallium lamp to energize thallium inside the cavity. The signature was obtained to establish the priority of his invention.

Around October of that same year, Charles Townes visited Bell Laboratories, where he was aiding work on masers, and lunched with Shawlow to whom he described his efforts to construct an infrared or optical maser. Shawlow was interested and Townes promised to give him a copy of his notes. The two men agreed to collaborate on this study and established that their work would form part of Townes' consultancy work with Bell Laboratories. Shawlow suggested removing all the walls of the resonant cavity, leaving only two which would form a Fabry–Pérot as a resonant cavity. As a student in Toronto, he had become familiar with the use of the Fabry–Pérot interferometer during research he carried out on hyperfine structure in atomic spectra under the direction of Professor Malcolm F Crawford. He later wrote:

'I had in mind from the beginning something like the Fabry–Pérot
interferometer I had used in my thesis studies. I realized, without even
having looked very carefully at the theory of this interferometer, that it
was a sort of resonator in that it would transmit some wavelengths
and reject others.'

In the notes Townes had given to Shawlow, he had performed some calculations using thallium atoms, which were excited using the ultraviolet light of a
thallium lamp. Such lamps were in use in Kusch's laboratory at Columbia
University for experiments on optical excitation of thallium atoms in an
atomic beam resonance experiment. Between 25 and 28 October, Townes
discussed with Gordon Gould, a student of Kusch who worked on the
atomic beam experiment, the properties of thallium lamps to find out how
much power they could be expected to deliver.

Shawlow quickly demonstrated that this thallium scheme of Townes
would not work easily and so they started to search for other materials. Eventually they chose potassium for the sole reason that it possessed lines in the
visible that Shawlow could measure with a spectrometer that he had bought
when he started to work on superconductors. This was the only optical
instrument he had. In the meantime Townes calculated the number of excited
atoms needed and performed some experiments. They then turned their
attention to the cavity to be used. In the end Shawlow realized that a wavelength selection could be made by considering the propagation directions of
the various wavelengths travelling in the cavity. With Townes they realized
that in order to travel back and forth between two plane parallel mirrors
the light should propagate along the axis of the system and that light emitted
in any other direction is quickly lost. On the basis of this reasoning they
decided to use a Fabry–Pérot cavity and on 29 January 1958 Townes
asked Solomon L Miller, a graduate student of Townes' working at Bell
Laboratories (who later went to IBM), to sign the notebook where these
considerations were described.

During the spring Shawlow and Townes decided to write up this work
for publication. It was customary at Bell Laboratories to circulate manu-
scripts among colleagues prior to publication in order to obtain technical
comments and improvements. A copy was also sent to Bell's patent office
to see whether it contained any invention worth patenting. As a result of
this process colleagues asked them to write more about the modes because
they did not believe that the Fabry–Pérot was able to select the required
wavelength and they required a calculation which Shawlow was at the time
not able to make. The patent office at first refused to patent either their
amplifier or their optical frequency oscillator because 'optical waves had
never been of any importance to communications and hence the invention
had little bearing on Bell System interests'. However, upon Townes'
insistence, a patent request was filed and it was delivered in March 1960.
The paper itself was received on 26 August 1958 and published in December

of the same year in *Physical Review*. Its authors were later awarded the Nobel prize; C H Townes in 1964, as we have seen, for his invention of the maser and proposal for the laser, and A L Shawlow in 1981 for a related subject: laser spectroscopy.

In their paper entitled 'Infrared and Optical Masers', Shawlow and Townes observed that, although it was possible in principle to extend maser techniques into the infrared and optical region to generate highly monochromatic and coherent radiation, a number of new aspects and problems arise which require both a quantitative re-orientation of the theoretical discussion and a considerable modification of the experimental technique used.

The declared purpose of the paper was to discuss theoretical aspects of maser-like devices for visible or infrared wavelengths and to outline the design considerations, so as to promote the realization of this new kind of maser, named by them as an optical maser (later to be called the laser, where 'L' stood for light). The principal points considered were the choice of the cavity and its mode selection properties, the expression of the gain of the device and some proposals on active materials.

Although one may expect that many materials could amplify, Townes observed that the excitation of atoms and molecules by means of light beams, electric discharges and other means, had been studied for years and nobody had observed amplification in the optical region. Therefore he assumed that obtaining amplification would be very difficult and all experiments should be planned with the greatest care. For this reason they chose to concentrate on gases of simple atoms, even though solid materials and molecules had certain advantages.

The most immediate problem was the realization of a resonant cavity. In the case of the maser an ordinary microwave cavity with metallic walls had been used. With a suitable design of such a cavity, it was possible to obtain just one resonant mode oscillating near the frequency corresponding to the radiative transition of the active system. In order to obtain such a single, isolated mode, the linear dimension of a cavity needs to be of the order of the wavelength which, at infrared frequencies, would be too small to be practical at this time. Hence, it was necessary to consider cavities whose dimensions were large compared with the wavelength and which could there-fore support a large number of modes within the frequency range of interest. Townes and Shawlow observed that it was necessary to find a way of select-ing only some of all these modes, otherwise the emission energy would be very small and would not be sufficient to overcome all the losses, exactly as occurs for a water-course that, if divided into thousands of streams, disperses and is absorbed by the ground without appearing to go anywhere. After some general considerations, the authors' choice was a Fabry–Pérot interferometer constructed from two perfectly reflecting plane parallel end walls. They showed that, thanks to the fact it was an open cavity and it was therefore deprived of lateral walls, being made of only two parallel mirrors, it trapped only the waves that were travelling parallel to the axis possessing wavelengths that were an integer factor of the cavity length (that is the distance between the mirrors). The problem is analogous to that which we discussed for the modes of the blackbody cavity in chapter 3. To extract light from such a cavity they suggested that one of the mirrors be partially transparent so as to allow a beam of light incident on it to be partially transmitted outside the cavity. The idea of adopting a cavity of dimensions much larger than the wavelength had been suggested not only due to the practical impossibility of constructing one of the correct dimensions for resonance, but also owing to the fact that the resonator should contain in its interior a reasonable quantity of active material.

Another problem treated in the paper was the determination of the minimum number of molecules or atoms of the active material which should be at the higher energy level to allow the generation of light by stimulated emission. This problem was simplified by considering a material with only two energy levels and the condition for oscillation was obtained by requiring that the power produced by stimulated emission be as great as that lost to the cavity walls or due to other types of absorption. When the two powers are equal, one says that the system is at the threshold for oscillation. A generated power just over threshold triggers stimulated emission.

Monochromaticity of a maser oscillator was also considered, and Shawlow and Townes observed that this property is very closely connected with the noise properties of the device as an amplifier. In the laser, noise arises from the spontaneous emission of radiation by the active material. They modified the calculation previously developed for masers and found that the linewidth was of the order of one millionth of the linewidth corresponding to spontaneous emission.

A paragraph in the paper was then devoted to discussing some specific example. Among gaseous systems they considered atomic potassium vapour pumped at 4047 Å, and caesium vapours. Shawlow had even performed some preliminary experiments on commercial potassium lamps and had asked Robert J Collins, a spectroscopist at Bell Laboratories, to measure the power output of these lamps. They calculated that, in the case of potassium vapours, a potassium lamp emitting about 1 mW at a wavelength of 4047 Å would suffice. This value seemed reasonable, considering that they had already succeeded in extracting about half a milliwatt from a small commercial potassium lamp which in ordinary conditions emitted only a tenth of a milliwatt at that wavelength. They also considered solid-state devices, although they were not very optimistic about these.

The work by Shawlow and Townes created considerable interest and many laboratories started to search for possible materials and methods for optical masers. Townes, with his group at Columbia, started efforts to build an optical maser with potassium. He worked with two graduate students Herman Z Cummins and Isaac Abella. At that same time Oliver S Heavens, then Professor of Physics at York University, York, England, who was then already a world expert on highly reflecting mirrors, joined this group. Townes had in fact realized that cavity mirrors were the most delicate part of the system and invited him for a sabbatical year. Their project utilized a long tube in which the potassium vapour was housed ready to be excited by means of an electrical discharge (as occurs in neon tubes used for shops signs). The cavity was created by putting the two mirrors inside the tube at its two extremities. The two mirrors needed to have a high reflectivity and to absorb very little light. This was obtained by depositing on them a series of layers of suitable materials with a technique in which Heavens was a master. Today we may ascribe the operational failure as being due to the degradation of the mirror coating owing to bombardment from the ions of the electrical discharge in the tube.

Shawlow, at Bell, had begun to consider ruby as a possible solid-state material but in 1959 he reached the conclusion that the energy levels later used by Maiman, were not suitable, and so he lost the opportunity to build one of the most popular lasers which exists today, notwithstanding he had correctly predicted that the structure of a solid state maser could be especially simple. In essence, it would be just a rod with one end totally reflecting and the other end nearly so. The sides would be left clear to admit pumping radiation.

CHAPTER 12

THE MISFORTUNE (OR FORTUNE?) OF GORDON GOULD

It is the opinion of historians of science and technology that, in general, it is an error to try to link an invention or a scientific discovery to a single individual or to a precise moment. An invention is a process that occurs over a time interval and to which usually many people contribute in a substantial manner. We saw this happen in the case of the invention of the maser and we will see this more so in the case of the laser. Shawlow and Townes were not, in fact, alone in studying how to extend the maser concept to the visible and infrared region and to predict the potential applications of an optical maser.

Gordon Gould, a Columbia student with the practical and intuitive mentality of an inventor and who focused entirely towards achieving the ultimate conquest of a patent, did not like to broadcast his ideas in the scientific literature yet had similar thoughts and, although he did not publish his results in the traditional way in scientific journals, created a legal claim with a series of requests for patents that formed the basis for a number of court cases which lasted for several years concerning the invention of the laser.

At 21 Gould earned a Bachelor in physics in 1941 from Union College, and in 1943 a Masters in optical spectroscopy at Yale University, where he learned how to use the Fabry–Pérot interferometer. When he completed his military service, he decided to dedicate himself to the inventing profession and found part-time work. He began by designing a type of contact lens and other items, including attempts to obtain synthetic diamond. However, in the end he concluded that to continue he needed a more solid scientific base, and in 1949 enrolled at Columbia University, where in 1951 he started his work for a PhD under Professor Polykarp Kusch. The thesis was concerned with the use of thallium molecular beams of which the excited energy levels were to be studied. By illuminating the thallium atoms with light from suitable lamps he first excited them into the desired level and then examined how they decayed from this state, with what efficiency the state was populated, and so on. However, the work progressed very slowly, so even in November 1957 Gould still had to write his thesis.

In fact, he had been interested in a project to build an optical maser, which he renamed the laser, by substituting the 'm' of maser that stands for

218

microwaves, with the 'l' that stands for light. When the first lasers were built, Bell Telephone did not like the name, refused to use it and did what they could to impose the name optical maser, clearly without success as the device came to be known as the laser.

The story that we may reconstruct from the depositions in the numerous trials that took place to claim the paternity of the invention, started in October 1957 when Gordon Gould, according to his own declarations was considering the possibility of using a device of the Fabry–Pérot type in a laser resonator. One day he received at home a telephone call from Townes whose office was near to Gould's on the tenth floor of the Physics Department Building at Columbia. Townes wanted some information on the very bright thallium lamps that Gould was employing for his thesis. Townes also in his notebook has registered his call. After this call, Gould became very excited and rushed to finish his studies as quickly as he could. On Friday 16 November 1957, Gould and his wife, a radiology assistant at Columbia, went to the proprietor of a candy store (who was a public notary and a friend of Gould's wife and of his family), who there and then put his seal on the first nine pages of Gould's laboratory notebook which contained the work 'Some rough calculations on the feasibility of laser light amplification by stimulated emission of radiation'.

In the notes, more than one hundred pages, he suggested containing the active substance in a tube 1 m long that was terminated with two reflecting mirrors (that is a typical Fabry–Pérot). He was considering also the possibility of placing the mirrors externally on the tube, by closing its extremities with two very homogeneous glass windows with an accurately worked planar surface with changes of thickness of the order of a fraction of wavelength—called optical facets—oriented at a particular angle, known as the Brewster angle. A glass slide, inclined with respect to the tube axis at the Brewster angle, allows light that has a particular direction of polarization to pass without any reflection; this light can therefore travel in and out and be reflected by mirrors connected externally to the tube without being attenuated, a very important property for a laser in which one must try to minimize the losses. Gould derived the conditions for maser oscillation, finding the right result, mentioned optical pumping as a possible excitation method, which he had discussed with Townes and, as a possible medium, the vapour of an alkali metal, quoting as an example potassium vapour, and then ruby or some rare earths. He quoted also pumping by collisions in a gaseous discharge, mentioning the helium–neon mixture as one of the possible gaseous media to excite. He then went on to discuss a large series of applications to spectrometry, interferometry, photochemistry, light amplification, radar, communications and nuclear fusion. In this list he spoke of frequency and length standards made with lasers, profile measurement systems, material treatment, techniques to make holes or cut materials and activation of chemical reactions, all made by means of laser light.

By an irony of fate it was Townes himself who introduced Gould to the use of a signed notebook as a method of establishing priority of claim to an invention! Gould's obsession for the laser had already cost him a great deal. His thesis advisor Professor Polykarp Kusch, according to Gould, would never let him substitute his thesis work for this subject, so in March 1958 Gould left Columbia without finishing the thesis and joined Technical Research Group (TRG) Inc.

TRG was one of those American companies that came about during the Cold War whose focus was defence and military contracts. It was founded as Technical Research Group in 1953 by three men who each had earned a doctorate, one in electronics, another in physics and the third in applied mathematics, and at first operated as a consulting agency. In 1955 laboratories and workshop were added. The principal work concerned antennae and radar, nuclear reactor physics and missile guidance, but it had also a small contract on masers and a programme on atomic frequency standards. Gould was assumed to work on this last project. Having not yet finished his thesis, TRG assured him some free time until July 1958 to complete his work. Gould, however, used this time not to work on his thesis but to work on his laser project.

He explored a great number of laser media and excitation methods: optical pumping, excitation by collisions with energetic electrons in a gas discharge, excitation by collisions among excited atoms and atoms of the active substance (collisions of the second kind). As media he was considering sodium vapours, sodium with mercury and helium, zinc and thallium excited by energy transfer from krypton and xenon, molecular iodine pumped with the light from a sodium lamp and europium sulphate optically excited in an aqueous solution. Many of his approaches were simple suggestions and he did not work in depth on the physics of the processes. Also many errors were present.

At the trial held on 8 December 1965 in Washington, Dr Alan Berman, a physicist friend and bridge partner of Gould, who was associate director at Hudson Laboratories Dobbs Ferry, Columbia, testified to a conversation he had with Gould in August 1958 at a beach party at Fire Island, New York. Gould at that time was with TRG and Berman said he had noted Gould was working on the laser to the detriment of his thesis. He was annoyed at him by observing he did not make any attempt to publish it in properly defined channels, as all physicists do, and added he thought this was an unscientific thing to do. It is worth noting that if Gould had followed Berman's advice he could have sent *Physical Review* a paper that could have been published simultaneously with the one by Shawlow and Townes!

In September 1958 Lawrence Goldmuntz, president of TRG, became aware that Gould was wasting his time working on a private project. Gould and Goldmuntz discussed the research on the laser and TRG took on Gould's project as its own. On 16 December 1958, Goldmuntz asked for $200 000 from the Aerojet-General Corporation of El Monte, California, which at that time owned 18% of TRG. A further $300 000 was also requested from the Advanced Research Projects Agency (ARPA) of the Pentagon to enable Gould's work on the laser to be used for optical radar, range finding and communication systems.

By coincidence Gould and Townes, at almost the same time, each had in their hands each other's work. Gould received a preprint of the Shawlow and Townes paper from Dr Maurice Newstein, a researcher at TRG: Townes, as an adviser to the federal government was reading the 200 page proposal submitted to ARPA. Other people too saw the latter and gave favourable reply to it: TRG received $998 000 for the project. The increase to the requested support was (and still is) very rare, not to say unique; however, the agency was heavily engaged in the problem of antiballistic missile defence, and because the laser, even though it did not yet exist, was one of the possible means one thought could be used, the agency asked TRG to study contemporarily the development of all kinds of proposed lasers and increased the support to $998 000, classifying the work. This was a great misfortune for Gould who was unable to work on this project at TRG because he had not got the necessary security clearance. In fact during the Second World War he was a member of a Marxist group and this was sufficient to deny him permission. As a consequence he could not lead the project, nor read the reports or participate directly in the experimental work. He was only the internal consultant of the research team.

On 6 April 1959 Gould and TRG filed a patent application for the laser in the United States, after which followed a series of requests for British patents which were granted. Shawlow and Townes had already filed their application in July 1958 and had the patent issued in March 1960. Gould and TRG appealed to the US Customs and Patent Appeals Court arguing that though their request had been made after that of Shawlow and Townes, Gould had the idea first. The principal proof was the notebook Gould had sealed on Friday 16 November 1957, after speaking with Townes. The trial was lost on 8 December 1965.

The Gould story does not end here. On 11 October 1977, after years of effort, he finally received his patent for an optically pumped laser amplifier. The patent arrived after 18 years of waiting: a record perhaps if one considers that three or four years is the usual waiting time! When the Gould patent went into operation those of Shawlow and Townes, issued in 1960, ended (in the US, patents at the time ran for 17 years). A second patent staking out three broad claims on laser applications was issued on 17 July 1979 to Gould by the United States Patent Office. Gould in the meantime had left TRG and given the management of his long-awaited patent to Refac Technology Development Corporation of New York which started a campaign to demand royalties, but the laser manufacturing companies, after paying royalties to Bell for nearly 20 years, had no intention of paying any money to Gould and another litigation began which eventually favoured Gould. The story went as follows.

The patents war

The laser-patent war raged for 30 years. Had Gould had been given good legal advice then, he could have applied for a patent before Townes and Shawlow even at the time of his notes in November 1957 and certainly would have been granted it. However, he thought he first had to realize his idea practically and so lost time and submitted his request only two years later, after that of Townes and Shawlow. However, as we have seen Gould did not surrender and with TRG received British patents for several different phases of laser technology. These patents never made Gould rich, but they did fuel his desire to keep his American claims alive. When Control Data Corp, which bought TRG in the early 1960s after liquidating the subsidiary's assets, Gould talked the company into allowing the contested patent rights to revert to him. In the meantime he remained active in the laser world as a professor at the Brooklyn Polytechnic Institute until 1973, when he left to found Optelecom, an early manufacturer of fibre optic data links.

The same year the US Court of Customs and Patent Appeals, in a suit over Q-switch (a technique to produce single laser pulses) patents, decided that the Shawlow–Townes patent did not adequately describe optical pumping of a laser medium. During the same period Gould decided to trade a half-interest in his pending laser patent to the Refac Technology Development Corp, a New York patent-licensing firm headed by Eugene Lang, in exchange for absorbing his legal costs.

So it was that in 1977 the Patent Office, 18 years after the original application was filed, issued Gould a patent on optical pumping of lasers (figure 49) and Refac immediately notified manufacturers that they would have to pay royalties on optically pumped lasers. Refac asked royalties ranging from 3.5% to 5% that generated more than $1 million per year in licensing fees in the sales of solid-state lasers alone, not to speak of other types of lasers that could have been covered by the patent. In its entire 17-year life, Townes' optical maser patent generated only $1 million in royalties because Bell according to an agreement with the government had agreed to ask only for low level royalties on its patents.

Therefore the laser producers resisted this request and Refac, barely a week after the patent was issued, filed a suit against Control Laser Corp, a leader in resisting the patent claims. Seven other laser companies joined Control Laser to fight the Gould patent.

The following year, Gould received a second patent on a broad range of laser applications including machining. News of that patent sent Refac's stock up $10 per share to $34, making Lang's 56% of the stock worth some $40 million. Speculators did not wait for the courts to act on the

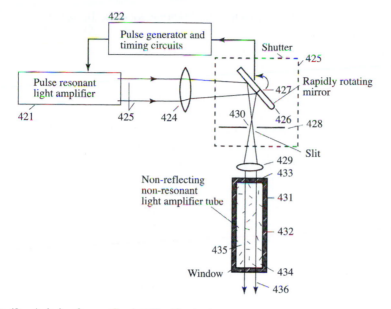

Figure 49. A design from a Gordon Gould patent.

patents. Gould sold part of his remaining interest to what seemed an unlikely buyer, the Panelrama Corp of Ardmore, PA, a public company that had operated retail building-support stores. The company changed its name to Patlex and bought 20% of Lerner David Littenberg and Samuel, a Westfield, New Jersey law firm that represented Gould's interests for a 25% stake in the patents.

One of the arguments that the laser companies brought to the courts in trying to resist Gould's patent was that the information he gave was insufficient to build a laser. So Gould at Optelecom Inc, in 1981, together with some other researchers and with the financial support of Patlex, employing a commercial sodium lamp manufactured by General Electric for use as a street light to pump an organic dye, rhodamine B, built a laser using only the design information disclosed in his 1959 patent application and other data readily available at that time. This result obviously dealt a blow for the companies that opposed the patent.

The first case to come to trial was against the tiny General Photonics. The company could not afford much defence, and on 1 March 1982 a federal judge ruled it owed royalties on the optical pumping patent. Soon afterward, however, patent opponents took advantage of new regulations that allowed the Patent Office to re-examine claims of patents that had already been issued. They were happy in early 1983, when the Patent Office denied Gould's claims. However, Gould was very stubborn and returned to court and succeeded in reinstating his claims. That proved the decisive victory, and long-delayed infringement cases finally came to trial. In May 1987, Patlex won its case against Cooper LaserSonics, a Palo Alto, CA, company formed by the merger of several smaller laser companies. In the same month, Lumonics settled three infringement suits out of court. Other companies settled with Patlex.

The final victory came in October 1987 in the long delayed Control Laser case. Days later, Gould received his third patent, on gas-discharge lasers. A fourth, on Brewster-angle windows for lasers, followed in 1988.

Soon everyone was talking with Patlex. The fate of the eight companies that initially fought the optical pumping patent shows how decisively Gould's forces won the patent war. Patlex bought the assets of General Photonics after it filed for bankruptcy and acquired Apollo Lasers in a stock exchange with Allied Chemical, which earlier had paid $10 million for the company. Hadron and National Research Group quietly faded away. Financial problems including court costs and licensing fees helped force the sale of Control Laser to Quantronix. Quantronix, in turn, fell victim to a market slump and was bought by Excel. Spectra-Physics has also gone through tough times, including a traumatic sale, but it remains a major player in the laser industry. Only Coherent remained an independent company.

Patlex itself merged in December 1992 with AutoFinance Group Inc, a financial-service company. Revenue from laser patents was about $7.5 million in the year ending 30 June 1992, and roughly $7 million for the following year. Gould received far more than he would have received if his first American patent had been issued promptly. That patent on optical pumping, expired on 11 August 1993. The applications patent expired in July 1996 and the discharge-pumping and Brewster-window patents will expire shortly.

To people who want to ask how much Townes' and Gould's ideas had developed from common information, considering that both were at Columbia and knew each other well, one may give an answer by making two considerations. The first is that an idea needs a receptive substrate to fully develop, that is all the general thoughts must be developed before retaining another person's idea and being able to work on it advantageously. In other words, ideas take root only in prepared minds. Even if talking to Townes had given Gould the idea that it is possible to excite atoms by optical pumping, Gould would already have developed the concept of using a population inversion, an optical cavity and so on in order to capitalize on them in his laser project.

Secondly, if we examine how the two men developed the laser idea and specifically how they went about solving one of the principal problems, namely the resonant cavity, we see that the two proposed solutions are typical of their two different personalities. Townes, the inventor of the maser, a microwave expert, started by considering a cube, that is the typical shape for a microwave cavity, in which the radiation travels by reflecting between the walls, and only later on Shawlow's suggestion removed all the walls except two. Gould, with an optical education, considered initially a cavity formed by a long tube (1 m) terminated by two plane parallel mirrors and then worked out all the possible configurations with plane external mirrors, curved mirrors, total reflecting prisms, and so on.

Gould was above all an inventor who, after having sketched in a note-book a draft of his idea, threw out a series of partially developed suggestions which were formalized in a proposal for a contract to assemble the device. Townes and Shawlow, with the minds of professional physicists, first thought about writing a paper to communicate their idea to the scientific world, not without having first patented it (do not forget they were supported by a commercial firm), and then worked theoretically on the details before dedicating themselves to experimental activity. There is therefore little doubt that also in this story the idea was born and developed independently and contemporarily by these three researchers.

CHAPTER 13

AND FINALLY—THE LASER!

Immediately after the publication of the paper by Shawlow and Townes, and even before this, a number of people had begun to think about different strategies for the production of inverted populations in the infrared and visible regions. The creative mentality of the researcher, who is driven to improve current knowledge and to push forwards its frontier in every possible direction without prejudice, led to the consideration of several different systems almost simultaneously and independently. In some cases, such as the one in which the radiation emitted via stimulated recombination of electron–hole pairs in semiconductors is utilized, the research preceded the discussion presented by Shawlow and Townes.

Of course the main themes were influenced by the ideas of these two researchers and most people were expecting the first laser action to take place in an excited gas. However, not everybody was working on gases. It so happened that the first working laser was realized in July 1960 at the Hughes Research Laboratories, Malibu (Southern California), by Theodore H Maiman, using ruby as the active material. There followed a huge number of other lasers in solid materials, gases and liquids, demonstrating that many people from different parts of the world had started to approach the problem from different angles whilst working more or less independently from one another. Furthermore, they showed how easy it was to make a laser, once the basic principles were understood.

Maiman begins construction of a ruby maser
Theodore H Maiman was born in 1927 and after studying at the University of Colorado, earning in 1955 a PhD in physics at Stanford University with a thesis on microwave spectroscopy, he was employed in industry initially as a researcher at Lockheed Aircraft, where he studied communication problems linked with guided missiles. Then he went to Hughes to work on the maser.

During his PhD work at Stanford, Maiman studied the fine structure of the excited states of helium using a measurement technique he had developed using a combination of electronics, microwaves and optics instrumentation.

At Hughes he went to work in the newly-formed Atomic Physics Department, where the principal interest was to generate higher coherent frequencies than were currently available. This was about the time that the ammonia maser came about. Hughes had an intense interest in maser research at that time. At first, however, Maiman worked on a different contract. When he finished this work he had wanted to work in a fundamental research capacity, but the Army Signal Corps, which sponsored the research, required at that time a practical X-band (that is at a wavelength of 3 cm) maser. They did not want any state of the art advances, they simply wanted that maser and Maiman was asked to head the project. He was not very enthusiastic at first because the project involved a practical device and he was more research-oriented. But then he became more interested and even though they had not demanded he make any tremendous advances he decided he could certainly make the maser more practical.

Masers at that time had two serious practical drawbacks. The main difficulty was that a solid state maser, which is the more useful type, needs to work at very low temperatures. Indeed, liquid helium temperature was needed, that is only 4 K higher than absolute zero. The other problem was that the conventional maser used a huge magnet, weighing about two tons, to obtain the Zeeman levels needed for the maser action. Inside this magnet was a dewar (a special container in which a liquid gas can be stored without it evaporating immediately) in which had to be poured liquid nitrogen in order to start lowering the temperature (nitrogen liquefies at $-166\,°C$). Inside this dewar, another one was placed which was full of liquid helium. The real maser was a small microwave cavity, with the crystal in its interior, which was positioned in the liquid helium dewar between the pole faces of the magnet. The magnet had to create a strong magnetic field within the whole volume occupied by the two dewars and the maser cavity, which justified its great size. The preferred maser material at that time was ruby. Maiman decided that there were a certain things he might be able to do still using ruby. He made a miniature cavity from ruby by cutting it into the shape of a small parallelepiped. Then he painted a highly conductive silver paint over the ruby and put a small hole in it. In this way the ruby behaved as both the active material and the resonant cavity and so space was saved. He then decided that instead of putting the double dewar inside the monster magnet, he could put a small permanent magnet inside the dewar. It was thought that the magnet would crack and break although it actually worked nicely. So the whole thing, magnet, dewars, and everything, was less than 15 kg instead of two tons, and it performed technically much better and was much more stable than before.

Later, he made an even smaller maser that weighed less than 2 kg and he developed a 'hot' maser cooled at liquid nitrogen temperature and then one cooled at only dry ice temperature.

The ruby laser

Speculations on laser materials during the first half of 1960 focused on gases, and more specifically on the vapours of alkaline metals excited by optical illumination, and noble gases excited by an electric discharge. The success obtained by Maiman with the ruby laser was truly a surprise. However, it was not an accidental discovery. Having already worked with ruby as a material for the microwave maser, Maiman employed it just as an initial reference material. In the beginning he performed some calculations but was discouraged because Irwin Wieder had published a paper which indicated that the quantum efficiency of ruby (that is the number of fluorescence light photons emitted for every absorbed light photon) was only around 1%.

Ruby is a crystal of aluminium oxide (Al_2O_3) in which a few chromium atoms are dispersed as impurities (we say that it is 'doped' with chromium). A chromium atom loses three of its electrons and becomes a chromium ion that replaces one of the aluminium ions in the lattice. These chromium ions have a series of energy levels in the visible region (figure 50) which provide the material, that otherwise would be a transparent crystal, with a colour between pink and dark red according to the quantity of chromium present. In the figure, two series of levels are shown so near to each other as to form practically two continuous bands. These two absorption bands are centred at wavelengths of 0.55 μm (green; this band is indicated by spectroscopes as $^4F_2$) and 0.42 μm (blue; this band is labelled as $^4F_1$) respectively. If the crystal is illuminated with green or blue light, the excited ions relax to two intermediate levels labelled with the letter ^2E in a very short time instead of decaying directly to the fundamental state. The transition from the green or blue band to these levels occurs without light emission but

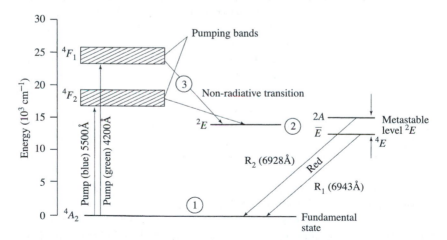

Figure 50. Energy levels of the chromium ion in ruby which are involved in laser emission.

gives excess energy to the lattice through collisions with its atoms. From these very close levels (called $2A$ and $\bar{E}$), the ions eventually decay slowly with a lifetime of the order of milliseconds to the fundamental level, this time by emitting red light that has a very narrow spectral distribution around 6928 Å (spectroscopists indicate this emission as the R_2 line) or 6943 Å (R_1 line) if the transition occurs from the higher level ($2A$) or from the lower one ($\bar{E}$) of the pair (doublet) ^2E respectively. The light emitted after the illumination of the crystal is called fluorescence. The names of the levels and bands were suggested by theoreticians according to considerations of group theory that reflected certain symmetry properties of the corresponding states that are of no interest here.

Irwin Wieder at Westinghouse Research Laboratories had studied how to obtain light corresponding to the narrow red lines of ruby, the R lines. Wieder used a tungsten lamp whose light was absorbed in order to excite the two green and blue bands of ruby. The energy was then transferred to the ^2E levels and Wieder calculated that the efficiency of this transfer was about 1% (that is about one hundredth of the energy absorbed in the two bands appeared as red light emitted by the R lines). If that was true, one red photon for every 100 absorbed photons practically ruled out the possibility of using optical pumping to obtain a laser. After having examined other materials, however, Maiman decided to perform more accurate measurements on ruby, by studying the spectroscopy of chromium ions in pink ruby for which he found that the quantum efficiency was indeed very high. These and other results obtained through a careful study of fluorescence formed the subject of a publication that was received by the American journal *Physical Review Letters* on 22 April 1960 and was published in the June of that same year.

In his research Maiman was helped only by an assistant, Irnee J D'Haenens, had only limited support from Hughes, and held the impression that a strong scepticism existed among his bosses George Birnbaum and Harold Lyons regarding the probability of his success.

Following his study published in *Physical Review Letters*, he had found the energy level distribution of the chromium ion that we have described and which is shown in figure 50 and had measured the lifetime of the ^2E level to be about 5 ms. This relatively long lifetime during which atoms remained in the metastable state and its successive decay with emission of radiation (radiative decay) are responsible for the fluorescence phenomenon in ruby, that is the phenomenon which gives this material its red colour. The rubies examined by Maiman were of the so-called pink species in which the chromium ion concentration is only about 0.05% by weight. Therefore, although the emission from the two lines at 6943 and 6928 Å is in the red, the overall colour of the crystal is pink (from which comes the name pink ruby). The measurement of the fluorescence quantum efficiency, that is the number of fluorescence photons emitted compared to the number of absorbed photons of the green exciting beam, had revealed a ratio close to unity; that is, for practically every green photon absorbed, there is one red photon emitted. This was a result completely at odds with that published by Wieder and one which made the possibility of making a laser feasible.

Maiman had calculated that a sufficiently intense green light pulse could populate the intermediate state ^2E in an appreciable way. Its consequence would be a variation in the fundamental state population. All these results encouraged him to use ruby for the first laser and to continue his preliminary calculations.

At this moment the principal problem was to find a green light source powerful enough to pump the atoms to the upper level. Roughly speaking, a lamp at high temperature emits light as if it was a blackbody. Preliminary calculations had shown that lamps equivalent to a blackbody at 5000 K were needed. Maiman started to make calculations with some commercial mercury lamps but found that the performances were just at the limit of usefulness. He then remembered that pulsed xenon flash lamps have an equivalent temperature of 8000 K. There was no reason which prevented the laser from working with a pulsed regime and indeed in many cases using pulsed sources was attractive.

We may now easily understand the dynamics of the process making recourse to figure 50. The illumination with green light excites some of the chromium ions from the fundamental level (in the figure indicated with the spectroscopic notation $^4A_2$ and with the number 1) to the band of levels denoted as $^4F_2$ or with the number 3. From here the ions decay rapidly in a fraction of a microsecond via collisions with the lattice atoms to the level ^2E (shown as 2), and from this level they return after about 5 ms to the fundamental level by emitting red light.

Maiman measured the decrease in the number of ions residing at the fundamental level produced by the absorption of green light, by observing the violet light at 4100 Å that is absorbed in the transition from $^4A_2$ to $^4F_1$, that is which raises the energy of the chromium ions from the fundamental level 1 to the band indicated with the letter $^4F_1$. When an intense pulse of radiation at 5600 Å was sent onto the sample, the radiation at 4100 Å, that at the same time was directed through it, suffered an abrupt increase that then decayed in about 5 ms. The effect is easily explained by observing that the pulse at 5600 Å, by exciting the ions from the fundamental level to the $^4F_2$ band, decreases the atoms available for excitation by the light at 4100 Å on the $^4F_1$ band and therefore decreases the absorption of blue light. Only after 5 ms, when the ions excited into the $^4F_2$ band after residing on the ^2E level return to the fundamental level, does the absorption of blue light come back to its usual value.

This experiment and others allowed Maiman to calculate that a change of population on the fundamental level of the order of 3% could be obtained.

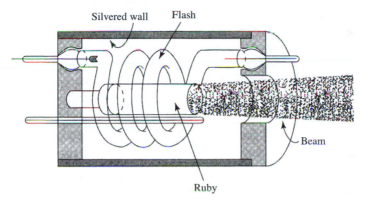

Figure 51. Layout of the Maiman ruby laser.

Encouraged by this result, he modified the experimental conditions to excite the greatest possible number of chromium ions from the fundamental level 1 to 2. For this he used a ruby cylinder encircled by a helical flash lamp. To direct most of the light onto the ruby sample he placed everything into a cylinder with silvered internal walls. In this way about 98% of the flash light was reflected onto the ruby cylinder. After carefully examining the catalogues he found out there were only three lamps that could work, all made by General Electric which produced a very powerful type of helical flash lamp used by professional photographers. They were listed under the names FT503, FT506 and FT624 (a monster designed for aerial photography). To obtain the resonant cavity he polished the two bases of the cylinder and made them parallel to each other. One was covered by an evaporated silver layer so as to be perfectly specular, the other one had a thinner silver layer so as to reflect about 60% of the red light emitted by the ruby and allow the rest to escape in order to be observed. The ruby cylinder was about 1 cm in diameter, about 2 cm in length and filled exactly the helices of the flash lamp (figure 51). Maiman had chosen the smallest of the three lamps, namely the FT506. During the flash, the energy of a capacitor bank charged up to several thousand volts was discharged. When the discharged energy was not very high, from the half silvered face of the ruby a red light flash was emitted that could be observed with the naked eye on a screen or could be detected with a suitable detector allowing one to observe how its intensity changed in time, revealing that it decayed with a characteristic time of about 5 ms, typical of the fluorescence. However, when the discharged energy reached some value (which in Maiman's experiment was about 0.7–1.0 joule) suddenly, viewing a screen in front of the half-silvered face, an intense red spot was observed of about 1 cm in diameter.

This result was obtained in May 1960, with a laser signal that was not very strong because the ruby used was a residue of the microwave maser experiments and was of poor optical quality. Maiman then ordered special rubies and prepared immediately a letter about his exciting results that he submitted on 24 June to *Physical Review Letters*. However, the journal editor did not publish the paper, considering that the maser physics had already reached a mature state and new progress did not deserve rapid publication. Needless to say, he had understood nothing about the matter. One must, however, not forget that the device was at the time labelled as an optical maser and also that people were inclined to believe Shawlow in that the R lines of ruby were not good for laser action and this may justify the editor's scepticism towards the exactness of the results. Anyhow Maiman made known his invention through a communication to the *New York Times* on 7 July 1960 and the paper refused by *Physical Review Letters* was eventually published in an English journal. The *Nature* issue of 6 August described the experiment.

When Hughes' public relations people took the photographs of the first laser, they used the FT503 flash lamp because it was more photogenic. So when the press release was circulated, everybody thought that this was the lamp used. There was a run on sales for those lamps and consequently all of the reproductions of the ruby laser made in other labs used the FT503.

There was not a great deal of enthusiasm at Hughes when Maiman worked on his project. In large companies very often there is a tremendous resistance to anything new and different. Many people were sceptical and believed optical masers could not be built; moreover they observed that many people were studying the problem without success, and finally even if a laser was obtained, how would it be useful? If that was not enough, Shawlow had said ruby could not work and Maiman was using just that material. People at Hughes questioned the money. Was it worth the company's investment to do this work? Maiman was not working on a contract but utilized general research funds. In any case at the end of nine months of work only $50 000 had been spent.

However, Maiman did not give up and was determined to continue, eventually obtaining a patent for his laser on 14 November 1967. Immediately after building his laser, he left Hughes and founded in 1962 a company, the Korad Corporation, which became a market leader in the construction of high power ruby lasers. In 1968 he sold Korad to Union Carbide and founded a venture capital firm called Maiman Associated and in 1972 the Laser Video Corporation. In 1977 he joined TRW Electronics in California where until 1983 he was vice-president of one section. He then became director of another company, the Pless Corporation Optronics. In 1984 he was inducted into the National Inventors Hall of Fame, was director of Control Laser Corporation, and sat on the Advisory Board of *Industrial Research Magazine*.

The day after Maiman's announcement that ruby was successful, many people continued to disbelieve it. In August, a group including Shawlow at Bell reproduced Maiman's laser and showed that it worked effectively, publishing the results in the October issue of *Physical Review Letters*. Many people who had not seen the English papers by Maiman read this paper and thought that Bell Laboratories were the first to build the laser. This misunderstanding came about because the proposal for a laser was made at Bell by Shawlow and Townes, who everybody knew were working on its practical implementation, while Hughes in California was completely extraneous to this research and isolated from the principal team based on the East Coast.

The operation of the laser is easily understood. When the flash lamp excitation is strong enough, the population of the ^2E state becomes larger than that of the fundamental state. In this situation, some of the spontaneously emitted fluorescence photons that travel parallel to the axis of the system and which are reflected back and forth between the reflecting extremities of the rod, passing many times through the amplifying medium, stimulate the emission of excited ions, producing thereby amplification through stimulated emission. The laser action is thus initiated by spontaneous emission and proceeds by amplifying only the radiation which, due to the selective properties of the cavity, travel back and forth along the axis of the rod. Photons travelling in any other direction not along the axis of the system are lost after a few reflections.

The laser action can be obtained on the R_1 or R_2 line. However, usually it occurs on R_1. The laser is characterized by certain properties peculiar to this kind of source: coherence, that is the ability to give interference phenomena; directionality and collimation; an emission on a very narrow band of frequencies with a very large power. The beam divergence, that is its angular aperture, was about $0.5°$, that is to say that after 10 m the beam had a spot less than 9 cm in diameter. Moreover the beam was spatially coherent, as was immediately demonstrated by observing its ability to produce interference fringes. The emitted power was about 10 kW, that is the light flux in the frequency range in which the emission occurs was nearly a million times that corresponding to sunlight at the Earth's surface in the same frequency range.

The laser output, examined using a detector coupled to an oscilloscope, appeared to be formed by a series of pulses each lasting a few microseconds (figure 52). This characteristic behaviour has been called spiking and a laser operating in this regime is said to work in free-generation. Very soon using a particular technique, called Q-switching or control of the merit factor, a method was found to force the laser to emit a single pulse much shorter in time. In this way light pulses that have instantaneous powers of hundreds or even thousands of megawatts can be obtained.

The construction of the ruby laser came as a bombshell in the scientific world, stimulating the development of a number of laser systems that nobody had believed possible a few months earlier. Practically any substance, air included, may be utilized to build a laser. In our story we will consider only a few cases. We will start with the construction of examples of solid

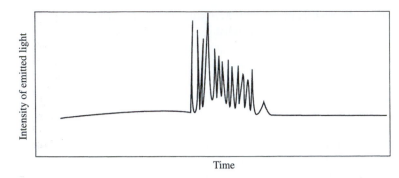

Figure 52. Ruby laser 'free generation' emission. The emission consists of individual oscillations (spikes).

state lasers and then we will describe the realization of the first gas laser, the helium–neon laser, which is even today one of the most commonly used lasers with excellent characteristics. We will examine the caesium laser, the neodymium laser, those based on organic dyes in liquid solutions, which can be tuned over a large frequency range and are some of the most versatile lasers today, and finally semiconductor lasers that are critically important in communication systems incorporating optical fibres, being for them the ideal source of radiation.

The second solid state laser

In September 1959 Townes organized a conference on 'Quantum Electronics—Resonance Phenomena' during which, even though no laser had yet been built, most of the informal discussions centred on lasers.

Peter Sorokin and Mirek J Stevenson of the IBM Thomas J Watson Research Center attended the conference and became enthusiasts of the laser concept. The Watson Center had been created in 1956 and offered all comforts to its researchers in pleasant countryside close to the cultural resources of New York.

The director of the physics section of the centre, William V Smith, after reading the paper by Shawlow and Townes, suggested that his microwave spectroscopy group, among whom were Sorokin and Stevenson, should redirect their effort towards lasers since he believed they would benefit IBM and help establish the reputation of the new laboratory.

Peter P Sorokin was the son of a sociology professor at Harvard University where he studied physics, receiving his PhD in 1958 with a dissertation on nuclear magnetic resonance under the supervision of Bloembergen. The young man had planned to go into theoretical solid state physics. In his second year of graduate work, together with a friend, he registered for a reading programme on nuclear magnetic resonance given by Bloembergen,

234

which they thought would be an easy course. At the time Bloembergen was not a very polished lecturer, and the two friends let everything go over their heads. At the end of the course, however, the professor announced he wanted a term paper from each student, to which Sorokin and his friend responded unsatisfactorily. Bloembergen commented 'These papers don't say anything about what I was teaching'. So Sorokin spent part of the summer trying to understand nuclear magnetic resonance, and then wrote another essay which this time Bloembergen accepted. By that time he felt he had invested so much time in the subject, which actually seemed interesting, that he may as well sign up to tackle a thesis with Bloembergen. First Bloembergen assigned him a theoretical problem, and for a year he sat at a desk with a pad of paper. Finally he came back to the professor and said: 'The divergent parts cancel, and all you have left are terms that are very hard to evaluate, but they are finite'. So Bloembergen looked at him and said 'Well, Peter, I think you had better do experiments'. So Sorokin received a thesis in which he had to make nuclear magnetic resonance measurements on caesium atoms. However, the caesium resonance appeared to have long relaxation times, which hindered the measurements. Another year went by and Sorokin became very discouraged; then, a young scientist came to the lab as a post-doctoral researcher and built a resonance system based on the crossed-coil approach of Bloch's group at Stanford. Sorokin understood immediately that this was the right way to perform the measurement, built a similar apparatus and finished his thesis.

After his thesis, Sorokin was hired by IBM to work on microwave resonance in solids. When the paper by Shawlow and Townes appeared, his boss suggested he study the possibility of building a laser. Along with Mirek Stevenson, who had obtained his PhD with Townes a couple of years before and who had been hired about the same time of Sorokin, he decided to focus on this new problem. Therefore, after the September 1959 Conference, they plunged into this work. They wanted to build a laser which emitted continuously using lamps having a power of the order of watts. Sorokin thought the principal problem was the pumping. To increase efficiency it was essential to decrease the losses and he considered eliminating the mirrors of the Fabry–Pérot cavity and substituting them with two total reflection prisms.

Total reflection is a phenomenon that occurs when light travels from a medium of larger refractive index to a second medium with smaller refractive index, for example from glass to air. If the angle the light beam in the glass makes with the normal to the glass–air surface is larger than a certain value (for a glass with $n = 1.5$, this angle is about 57°) the light is fully reflected and does not transmit into the air. In this way, losses due to absorption in the mirror material are eliminated. The head of the physics section of the centre, William V Smith, suggested choosing a crystal with a refractive index just equal to the value needed to obtain total reflection of the rays which hit the prism, so the effect could be used to select modes. Accordingly

Sorokin chose a crystal of calcium fluoride that possessed just the required refractive index value. The problem now was to find a material to put inside as the laser emission medium. After examination of the scientific publications on the subject, Sorokin found that the Russian P P Feofilov had studied light emission from uranium or samarium ions in calcium fluoride. Uranium has a fluorescence emission at about 2.5 μm. The uranium or samarium ions in the calcium fluoride substitute calcium ions in the crystal and have energy levels similar to those of chromium in ruby with the only difference (shown in figure 53) being that there is one additional level and therefore light emission may occur between a level populated by the decay of one band to an intermediate level which, if one works at low temperature, is practically unpopulated because thermal agitation is not able to populate it from the ground state. This circumstance, which we may describe in terms of a four level system, allows population inversion to be more easily obtained. Calcium fluoride doped with uranium has moreover a strong absorption band in the visible that could be pumped with a high pressure

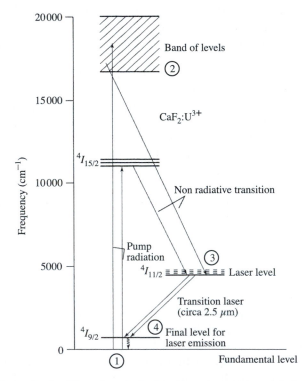

Figure 53. Energy levels of three times ionized uranium atoms in a crystal of calcium fluoride ($CaF_2 : U^{3+}$). The transition is obtained by pumping from the ground level (1) to the band (2). (The light absorbed by levels $^4I_{15/2}$ can be neglected.) Electrons decay from band (2) to levels (3) and the laser transition at about 2.05 μm occurs between (3) and (4).

xenon arc lamp. The system needed, however, to be cooled at a low temperature.

The two researchers asked a specialist firm to provide crystals doped with uranium and samarium and, a while after receiving the crystals, heard about the success obtained by Maiman with ruby. They immediately decided to forget about the idea of making a resonant cavity using total reflection, cut their calcium fluoride crystals into the shape of cylinders and silvered the two extremities. They bought beautiful flash lamps and early in November obtained the laser effect from the crystal doped with uranium and a short while afterwards with that doped with samarium.

After succeeding in making uranium lase, they wrote a paper for *Physical Review Letters*. Stevenson, being direct and aggressive, said: 'We're not going to send it. We're going to drive down to Brookhaven and tell Sam Goudsmit [the editor] we want a decision before we leave.' Sorokin said 'Mirek, you can't do that.' 'Nope, we're going to do that.' So they went to see Goudsmit. He was slightly confused about the difference between masers and lasers, and said he did not want another 'maser' paper just as he had done with the Maiman's paper. But Stevenson insisted so much he finally accepted it for publication and as they were leaving, Goudsmit said 'Next time, tell your people from IBM not to come down here with machine guns'.

Due to the operation being based on four levels, instead of three as for ruby, the laser worked with pump powers that were ten times lower than required for ruby. Uranium has a strong absorption band in the green-blue region. The laser oscillation corresponded to a wavelength in the infrared at 2.49 µm. The laser mounting was similar to that for ruby, only more complex because the crystal had to be cooled to liquid helium temperatures and therefore had to be placed inside a dewar.

Some time later, Sorokin, together with a technician of his, John Lankard, built another kind of laser with a liquid that was the first of a series of lasers, of which we will speak later, that utilize organic dyes dissolved in liquids, developed in a number of laboratories around the world and still very much used today.

The helium–neon laser

Besides Shawlow, two more researchers at Bell Laboratories started work on lasers in 1958: Ali Javan and John H Sanders. Javan was a native of Iran who received his PhD with Townes in 1954 for microwave spectroscopy work and remained in Townes' group for four years working on microwave spectroscopy and masers. After his thesis, while Townes was on sabbatical leave in Paris and Tokyo, Javan became more involved with masers and had an idea for a three-level maser before a group from Bell Laboratories published the first experimental work on the subject. He found a method to achieve amplification without population inversion using a particular effect known as the Raman effect in a three-level system. However, he published his results after the Bell group in 1957 and 1958.

In April 1958, while looking for a position at Bell, he spoke with Art Shawlow who told him about lasers. In August 1958 he was hired at Bell and in October started systematic studies on lasers. Initially at Bell he had an ethical dilemma. RCA had previously inspected his notebooks on the three-level maser and had established that his dates preceded the Bell dates. They paid him $1000 for patent rights and were contesting Bell Laboratories' application. For the first six months or so working at Bell, Javan was dealing with RCA and Bell patent attorneys. Luckily, RCA carried out marketing studies and concluded that the maser amplifier was not commercially viable and so agreed to drop their case and let Bell have the patent. In this way, the litigation that would have been very hard fought, considering the economical power of Bell, did not occur.

So eventually Javan could dedicate himself to the idea of making a laser, that he was thinking of building using gases, publishing the proposal in a letter to *Physical Review Letters* in 1959. He had decided to use a gas as an active medium because he believed the simplicity of gases would simplify the study. However, he thought it was not possible to have sufficiently powerful lamps to pump directly the atoms into the excited states and therefore considered excitations either by direct electron collisions with pure neon as a medium, or by collisions of a second kind. In the latter approach, the discharge tube could be filled with two gases chosen in such a way that the atoms of the first gas, excited by collisions with the electrons of an electric discharge, could transfer their energy to atoms of the second gas by exciting them. Several gaseous mixtures had an energy level structure that satisfied this request. In fact it is necessary that the energy level of the second gas has an energy practically equal to the excitation energy of the first gas. Among the possible combinations, Javan chose the combination of helium and neon gases whose levels are shown in figure 54. He concluded that any single physical process tends to produce a Boltzmann distribution of the different energy levels (that is, it populates the lower level more than the upper one). Therefore a medium with population inversion can be produced in a steady-state process only as a result of the competition of several physical processes proceeding at different rates.

These considerations can be better understood by considering for example a tree with several branches (two in figure 55) onto which we send monkeys. Let us consider populating initially the branches according to Boltzmann statistics, that is say four monkeys on the upper branch (1), five on the lower one (2) and six at ground (fundamental level 3). From the three levels, the fundamental one is more populated and the higher branches are increasingly less populated. The monkeys, however, are not at rest and drop from the branches (we may think one every minute for

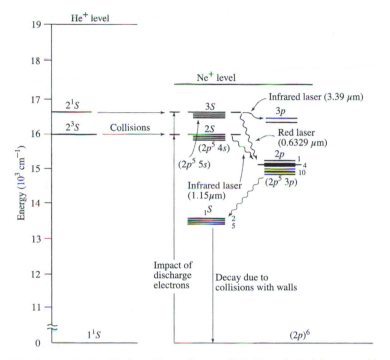

Figure 54. Energy levels of helium (He) and neon (Ne). The principal laser transitions are shown.

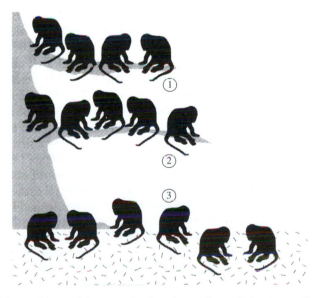

Figure 55. The monkeys on the tree are distributed according to Boltzmann statistics. There are more at ground and fewer on the higher branches.

239

example) so every minute we replace one. In this way the population on the levels remains the same in time (we are in an equilibrium situation). Suppose now we continue to populate the branches at this rate (one monkey per minute) but at the same time we wet the branch 2 so as to make it slippery. Now the monkeys are no more able to remain on this branch for more than 10 s, for example, and therefore the branch depopulates rapidly and soon there are more monkeys on branch 1 than on 2, that is the population between these two branches has inverted, by using the fact that the residence time of monkeys on the different branches is different. Although this is a very simplified vision, it helps to understand Javan's reasoning.

The choice of the helium–neon mixture arose through a careful selection based on obtaining a system with the promise of being the optimum medium, and the ensuing success of this laser gave *a posteriori* full credit to Javan. Even after he had convinced himself that helium–neon was the best gas medium, there were many non-believers who told him that the gas discharges were too chaotic. They said there were many uncertainties and that his efforts amounted to a wild goose chase.

Javan was spending a lot of money but fortunately the system worked otherwise the administrators at the end of the project would surely have shut down the experiment. In the end, Bell admitted to having spent two million dollars on this research and even though probably this figure is perhaps exaggerated, certainly the project required considerable financial effort.

Meanwhile, John Sanders—an experimental physicist of Oxford University—had been invited to Bell from January to September 1959 by Rudolph Kompfner, an associate director of the division on communication science, to try to realize an infrared maser. With less than one year of time for the research, Sanders did not lose any time in refined theoretical studies and decided to excite pure helium in a discharge tube inside a Fabry–Pérot cavity and attempted to obtain the laser effect through a process of trial and error by varying the discharge parameters. The maximum distance at which the Fabry–Pérot mirrors could be put and still remain aligned parallel to each other so as to realize the cavity was 15 cm and Sanders had to use discharge tubes no longer than this.

Javan judged this was a fundamental limitation. He expected the gas gain was so small that Sanders' cavity would never work. The tube used by Javan had to be much longer and because it was not possible to align the mirrors of the Fabry–Pérot over such a distance, he decided to determine first the required values of the parameters to obtain a working device and then to try to align the mirrors by trial and error. That is how it worked. Without all that preliminary work to preselect the HeNe discharge at a known gain, it would have been impossible to make it work.

Sanders sent a letter to *Physical Review Letters* in which he observed that it was difficult to obtain a sufficient number of excited atoms with a flash lamp, and suggested the use of excitation produced by electron collisions. Such an excitation could easily be produced in an electrical discharge in a gas or a vapour. Population inversion could be produced if excited states with a long lifetime existed in the active material and if there were states present at lower energies with a short lifetime, as we said when speaking of monkeys.

The very next paper in the same issue of *Physical Review Letters* was written by Ali Javan who too had considered these problems and among the schemes he proposed was a very ingenious one. Let us consider a long-lived state of a gas. During the discharge this state can be populated in an appreciable way due to its long lifetime. If now the excited state of a second atom has an energy very near to that of the long-lived state, it is very probable that in a collision energy will be transferred from the first to the second atom that therefore becomes excited. If this atom has other states at lower energy, these are not excited and therefore inversion of population can be realized between the higher energy state and the lower energy ones. In his work Javan mentioned mixtures of krypton and mercury and helium and neon. The paper was received at *Physical Review Letters* on 3 June 1959.

Javan worked in strict contact with William R Bennett Jr, a spectroscopist at Yale University who had been his friend at Columbia. The two men worked until late at night for a full year. In the spring of 1959, Javan asked Donald R Herriott, a specialist at Bell on optical apparatus, to collaborate with them on the project. One of the principal problems was the sealing of the discharge tube with two transparent windows of very good optical quality so as not to distort the output beam. Then one had to position the mirrors. The structure was designed (figure 56) with the mirrors inside the discharge tube providing a special mounting with micrometric

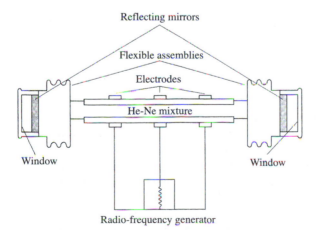

Figure 56. Layout of the helium–neon laser built by Javan, Bennett and Herriott.

screws which facilitated the alignment. In September 1959 Bennett left Yale to join Bell and together with Javan started an intensive and meticulous programme to calculate and measure the spectroscopic properties of helium–neon mixtures under various conditions in order to determine the factors governing population inversion. They found that in the best conditions one may expect only a very low gain, of the order of 1.5%. Such a low gain made it absolutely necessary to minimize losses and to employ mirrors with the highest possible reflectivity. These were mirrors obtained by depositing many layers of suitable dielectric materials one on top of the other, on a transparent support, and eventually the three men were able to use mirrors which at the wavelength of 1.15 μm had a reflectivity of 98.9%.

In 1960 Javan, Bennett and Herriott finally tested their laser. Initially they attempted to generate an electric discharge in a quartz tube that contained the gas mixture using a powerful magnetron, but the tube melted. The apparatus had to be remade and after other setbacks, on 12 December 1960, they realized the discharge in the new tube and tried to align the mirrors to obtain the lasing effect but without success. Then in the afternoon, Herriott saw the signal: 'I was casually turning the micrometer on one of the mirror adjustments when a signal suddenly appeared on the oscilloscope. We adjusted the monochrometer and found the signal peaked at 1.153 microns, one of the expected wavelengths'. The first laser using a gas as an operating medium was born! The emitted radiation was in the near infrared and was therefore invisible to the eye and had to be revealed with a suitable detector connected to an oscilloscope.

About six months earlier Ed Ballik, a technician who helped greatly in the work and later obtained a degree at Oxford University and went on to teach in Canada, had brought in a bottle of wine that was a hundred years old. They kept it for opening after the laser had been shown to work. A few days later, Javan called the head of Bell Laboratories and invited him to come and taste this hundred year old wine. He said he would be very glad to come, then said. 'Oh, oh, Ali. We have a problem!' This was two or three in the afternoon. He wouldn't tell Javan what the problem was, but he said he would come at 5.30. Later that afternoon, a memo was circulated through the lab. It turned out that some months earlier they had disallowed alcohol on the premises. The new memo stated no alcohol was permitted unless it was over 100 years old. After that he toasted their success without infringing any regulations!

The first laser operation took place at the transition at 1.15 μm, in the near infrared. Javan, in fact, had used mirrors that had maximum reflectivity at that wavelength, which corresponded to one of the possible transitions. He was aware there were also other possible wavelengths. At first he chose that particular wavelength because his study showed that for that emission one could expect higher gain. To use transitions in the visible region, the bore of the discharge tube needed to be so small that it would have been impossible to align the plane mirrors that were used at the time to make the Fabry–Pérot cavity.

The laser assembled by Javan was a discharge tube containing neon and helium at pressures of 0.1 and 1 torr respectively (1 torr is nearly one thousandth part of the pressure of one atmosphere). The quartz tube was 80 cm long with an inside diameter of 1.5 cm. At each end of the tube there was a metal chamber containing high reflection plane mirrors. Flexible bellows were used in the end chambers to allow external mechanical tuning of the Fabry–Pérot plates. This enabled the reflectors to be aligned for parallelism within six seconds of arc. Two flat windows optically polished so that their surface roughness did not vary to an extent greater than 100 Å were provided at the ends of the system which allowed the laser beam to be transmitted without any distortion. The electric discharge was excited by means of external electrodes using a 28 MHz generator. The input power was around 50 W. The high reflectivity of the mirrors was obtained by evaporating 13 layers of dielectric films. Reflectivity was 98.9% between 1.1 and 1.2 μm. The laser worked in continuous operation and was the first of this kind.

Following the example of Hughes, Bell also gave a public demonstration of the helium–neon laser on 14 December 1960 and in order to demonstrate the importance it could have for communications, a transmission of a telephone conversation was organized utilizing as a carrier the laser beam that was modulated by the telephonic signal.

The HeNe laser was so-called by employing the chemical symbol of its constituents. It was presented to the press on 31 January 1961. The paper describing the laser was received by *Physical Review Letters* on 30 December 1960. Soon after, Javan joined the faculty of the Massachusetts Institute of Technology.

While Javan was working on his experiments in the spring of 1960 two researchers at Bell started to study if modes exist in the Fabry–Pérot cavity. It was known that modes exist in microwave cavities, but there was no definite proof they existed in a Fabry–Pérot that is very different from the closed box of a traditional microwave cavity. They found that modes indeed existed and their results stimulated other Bell researchers, Gary D Boyd, James Gordon and Herwig Kogelnik, to find analytical solutions in the case of mirrors of spherical shape. The importance of the study of resonant cavities in promoting the development of gas lasers cannot be underestimated. Until this theoretical work was done, the gas laser was at best a marginally operating device whose oscillation depended upon almost impossible tolerances in the end mirror adjustments. Theoretical studies on curved mirror resonators showed that resonators could be devised that were relatively insensitive to mirror adjustment and whose intrinsic losses could be lower than those of a plane-parallel resonator, allowing observation of laser action in media with much lower gain than was at that time thought possible. The plane-parallel

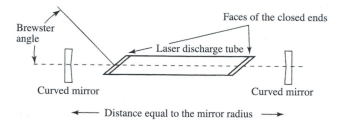

Figure 57. Confocal optical resonator. The discharge tube in which the excited gas is contained is closed by facets inclined at the Brewster angle. The curved mirrors of equal radius are located externally with a distance between them equal to their curvature radius.

resonator has since almost dropped out of existence for practical laser work, and all discoveries of new gas laser transitions have been performed with curved mirror resonators.

In 1961 Bell Laboratories started a large laser research programme and scientists previously interested in other problems were redirected to work on lasers, while new staff were hired. The solution to use identical spherical mirrors each positioned at the focal point of its partner (this arrangement is known as confocal mirrors) suggested how the building difficulties of the first Javan laser could be overcome. As a result, William W Rigrod, Herwig Kogelnik, Donald R Herriott and D J Brangaccio, in the spring of 1962, built the first confocal cavity using two spherical mirrors that concentrated light along the axis of the tube, and which were placed externally to the discharge tube, and obtained emission of the red line at 6328 Å. Some light is inevitably lost through the end windows. These losses could, however, be minimized by orienting the windows under a particular angle, called the Brewster angle, as we already described. The new laser mounting is shown in figure 57.

Besides providing great stimulus for applications, this visible laser brought an improved understanding of the problems connected with the optimization of laser structures and defined precisely the difference between a coherent light beam, as is emitted by a laser, and an incoherent beam such as that emitted by any normal lamp. The work also stimulated the development of other types of lasers that were invented in great numbers during the following years.

The helium–neon laser today can be made to oscillate on one of the possible transitions shown in the figure, by using multilayer dielectric mirrors optimized to have maximal reflection at the desired wavelength. The possible wavelengths are 3.39 µm, 1.153 µm and 6328 Å. The visible laser is the most popular today. Such lasers are commercially available with powers between 0.5 and 50 mW. More recently lasers have been developed in the green at 5435 Å, yellow at 5941 Å, orange at 6120 Å and in the near infrared at 1.523 µm.

The caesium laser

The year 1961 witnessed the realization of two more lasers on which people had worked since the very beginning of the laser concept. One of them was the caesium laser. Townes and Shawlow, after writing their paper, had agreed that Townes would attempt to build a laser with potassium vapours, both because the calculations had shown it could work but also because potassium is a simple monatomic gas with well known properties and he desired a system whose properties could be analysed in detail. 'My style of physics', he said later, 'has always been to think through a problem theoretically, analyse it, and then do an experiment which has to work. If it doesn't work at first, you make it work. You analyse and duplicate the theoretical conditions in the laboratory until you beat the problem into submission.' His preliminary calculations had revealed that a potassium laser would have a highly monochromatic output that would have been very useful for special applications. It would also have had drawbacks: low efficiency (about 0.1%) and a power output of fractions of a milliwatt.

While Townes concentrated on potassium vapours, Shawlow at Bell was studying ruby, concluding, however, that the lines later used by Maiman to build the first laser were not suitable. Townes, after asking for and obtaining financial support from the Air Force Office of Scientific Research, recruited two graduate students to work on the project: Herman Z Cummins and Isaac D Abella.

However, the project encountered a series of problems. The potassium vapours were darkening the glass tube of the discharge and chemically attacking the seals. Often the gas blew up the distillation apparatus; at other times, it picked up impurities during distillation such that, later, potassium ions that had been excited would lose energy. At the end of 1959, Townes asked Oliver S Heavens, a British expert in the physics of dielectric layer mirrors, to come to help in the research and had decided to shift to a laser which used caesium vapour instead of potassium, pumping them with a helium lamp.

One of the narrow absorption lines of the caesium atom has exactly the same energy as one of the narrow helium lines. It is therefore possible to use the light emitted at this wavelength (389 nm) from a helium lamp to pump selectively a caesium level which populates more than the lower levels. So population inversion is obtained (figure 58). After the announcement by Maiman, Townes shifted Abella to work on ruby, while Cummins continued on the caesium project. The caesium laser was made to work at TRG between the end of 1961 and the first months of 1962 by Paul Rabinowitz, Stephen F Jacobs and Gould and emitted radiation at 3.20 and 7.18 μm. It was one of the lasers made possible using confocal mirrors. Gould had mentioned in his patent application that it was a feasible but not a promising laser. The research at TRG had started with potassium, but after a seminar by Oliver Heavens who had spoken of caesium as a better candidate, TRG also

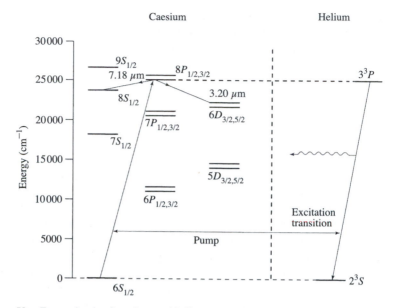

Figure 58. Energy levels of caesium and helium.

diverted to study this material and, pushed by the necessity to show the government that the million dollars they had received had been well spent, eventually succeeded in obtaining population inversion in March 1961 and oscillation early in 1962.

This laser was more a curiosity than a practical source of radiation because today there are other more easily constructed lasers which emit in the same wavelength range, and in addition caesium vapour is noxious.

The neodymium laser

The other laser realized in 1961, and still today one of the most important, is the neodymium glass laser. In 1959–60, the American Optical Company also entered laser research, through the work of one of its scientists, Elias Snitzer. American Optical's original focus was optical instruments and ophthalmic products. It had also strong capabilities in glassmaking and glassworking. During the 1950s, the company decided to expand its product lines, and it therefore initiated research projects in new areas such as military electro-optics and fibre optics. Elias Snitzer was hired into the research group in early 1959, and began with research on the propagation of electromagnetic waves through optical fibres. For the company, this work would strengthen its already considerable patent position in fibre optics and bolster its image in the field within the general scientific community. Snitzer saw that there were connections between his optical fibre research and laser work. Since a glass fibre may sustain electromagnetic modes, it could be converted into a laser resonator if one placed mirrors at its ends. This prospect was the more interesting because of the doubts within the scientific community about whether the Fabry–Pérot resonator that was being used with gaseous media would work. The glass itself could become a lasing material by doping it with a suitable substance like samarium or ytterbium that could possess the required levels for incoherent light excitation sent through the sides or the end of the fibre. Snitzer considered he would be able to concentrate even more pumping light into the fibre by covering it with a thick layer of glass with a slightly different refractive index.

Early in 1960, Snitzer with two more colleagues began examining a series of glass fibres doped with ions that had fluorescent lines in the visible range. Glass was an unusual choice. All the materials under study were gases or crystals. After Maiman's successful laser, Snitzer also tried with ruby fibres. Until then he had used high pressure mercury lamps which emitted light in a continuous way. Now he gave up and bought flash lamps. The group examined 200 fibres. At the end of 1960 Snitzer's two co-workers had transferred to an Air Force classified project aimed at creating a solar-powered laser transmitter. Snitzer continued alone and decided to shift from the visible to the infrared. This decision meant changing the doping materials. In the infrared, rare earths like neodymium, praseodymium, holmium, erbium and thallium could be used. Snitzer decided also to give up with fibres and concentrate on a simple doped glass rod. In October 1961 he succeeded in obtaining the laser effect from a neodymium-doped glass rod.

The neodymium ions exhibit narrow spectra both when they are incorporated in crystals and when in amorphous materials like glass. There are several advantages in using glass as a laser material. The preparation methods are well established and making glass is certainly easier than growing crystals. Besides, the homogeneity that can be obtained with glasses is much better than with crystals and therefore much larger samples can be produced. Moreover glasses doped with rare earth ions had already been in production for many years for their application as photographic lenses.

The levels of neodymium in glass are shown in figure 59. The $^4F_{3/2}$ level is the fluorescent one and laser transitions occur between this level and the $^4I_{13/2}$, $^4I_{11/2}$ and $^4I_{9/2}$ levels at wavelengths of 0.88, 1.06 and 1.35 μm respectively. Excitation is obtained by optically pumping from the fundamental level to levels above the $^4F_{3/2}$ state. There are three levels which absorb infrared radiation, levels which absorb in the yellow region around 5800 Å and other levels that absorb mostly in the ultraviolet. In the figure the absorbing levels higher than the $^4F_{3/2}$ level are shown as bold lines. From these levels the excited atoms decay with non-radiative transitions to the $^4F_{3/2}$ level from which the laser emissions begin.

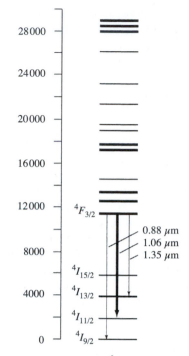

Figure 59.   Energy levels of neodymium ions (Nd^{3+}) dispersed in a Barium 'crown' glass.

Lasers using neodymium in glass are important first of all because they represent an example of a solid material different from a synthetic crystal, also because certain glasses doped with neodymium have great output energies per unit volume, and finally because the versatility of the glass matrix allows lasers to be shaped into rods or fibres.

In the same year the laser effect was obtained by L F Johnson, G D Boyd, K Nassau and R R Soden at Bell Telephone Laboratories with neodymium doped into a crystal of calcium tungstate. Their laser operated continuously and was cooled at liquid nitrogen temperature. The emission was at 1.06 µm.

In December 1961 ARPA (Advanced Research Projects Agency, by 1959 became the agency for the support of basic research into advanced military technology) organized a scientific committee who gave high priority to research on ruby and glass lasers. The following year, Snitzer obtained emission from a fibre of 32 µm diameter. Today rare earth ions incorporated into glass fibres make excellent amplifiers for signals sent along optical fibres and used for telecommunications.

The trivalent ion of neodymium has been incorporated in a great number of different hosts. One of those is a lattice of $Y_3Al_5O_{12}$, which is commonly represented by the YAG acronym for yttrium aluminium

248

garnet. The neodymium energy levels are essentially the same, irrespective of the host matrix. The neodymium YAG laser is used both in continuous and pulsed operation. This laser was made at Bell Laboratories by Joseph E Geusic, who in 1962 with E D Scovil started to write a review paper on masers and lasers in which an analogy was discussed that Scovil had made between an optically pumped laser and a thermodynamic heat pump. The analogy yielded to certain criteria for selecting laser materials which guided Geusic to select about 40 crystals among which was YAG. The trouble with this material was that there were no sufficiently long crystals. Geusic asked Bell colleagues, among whom were LeGrand G van Uitert, how to obtain such crystals. The two men at first obtained from an outside company a sample on which they performed encouraging measurements. Then van Uitert succeeded in growing a long enough crystal of good optical quality, and Geusic and his technician Horatio M Marcos had the laser working in a continuous way showing that it required a power about one fifth of that used for calcium tungstate with neodymium.

Geusic and van Uitert started a collaboration with the Linde division of Union Carbide to develop longer crystals of better optical quality and demonstrated all the advantages of this laser that constitutes a valid alternative to other power lasers (ruby and CO_2). This laser is a typical example of how multidisciplinary collaboration, characteristic of the large American laboratories, permitted in a couple of years the development of a new laser with exceptional performance.

Organic dye lasers

If most of the lasers we have considered up to this point arose as a result of a highly coordinated effort, and required the employment of advanced technologies, explaining in some way why they were all developed in the United States, the case of organic dyes (simple dyes) can be considered to be different. The first laser of this kind arose by chance, thanks to the technique, called Q-switching, introduced by Robert W Hellwarth at the Hughes Research Laboratories, Malibu, California, in 1961. A method of extracting great power through a single short pulse from a ruby laser consists of keeping the losses of the resonant cavity high so that the laser action does not start until the pump light has populated the upper laser level to its maximum extent. If at that very moment losses are abruptly lowered a cascade starts by stimulated emission that discharges all the energy of the excited levels in a single shot. In this way it is possible to obtain from a ruby laser a single pulse which lasts typically 30 ns and has power of the order of tens or hundreds of millions watts (megawatt). Initially, the method of achieving high losses was very crude. Between the two mirrors that formed the cavity an opaque disc was inserted that had a small slit and rotated at high speed. The pumping flash was powered at the right time such that when the population on the upper level was a maximum, the slit was aligned with the mirrors so that the cavity presented minimum losses. However, the system was too slow and was not easy to synchronize. Often, instead of one high power pulse, two or three pulses were obtained at much lower power.

So people began thinking of other methods. Among these was one self-regulating. When light is incident on an absorbing substance, it is absorbed because molecules that are in the lower energy state are excited to upper levels. If the light intensity is very high, however, most of molecules in the lower energy state are brought to the upper state and now the light can be transmitted unperturbed. The absorbing material becomes 'bleached' or 'saturated' as one usually says.

At IBM, Peter Sorokin and John Lankard showed in 1966 that a simple and convenient material to use for this effect could be a group of organic dyes called metallic phthalocyanines dissolved in organic liquids. Phthalocyanines are complexes with metal ions at their centres. The two researchers asked a colleague of theirs, John Luzzi, to synthesize some and Luzzi, rather than making a gram which would have been the norm, made them a whole pound. So Sorokin placed these phthalocyanines right in the cavity of a ruby laser and switched on the laser. A single powerful pulse, 20 ns long, was the immediate result.

Whilst trying to better understand what was happening, Sorokin thought that these materials could also be used for other experiments and focused on two of them. In one experiment he wanted to stimulate an effect known today as Raman scattering or the Raman effect, discovered in 1928 by the Indian physicist Chandrasekhara Venkata Raman (1888–1970), who in 1930 was awarded the physics Nobel prize. Raman had shown, that under certain conditions, a fraction of the light sent onto a material is absorbed and re-emitted at a slightly different frequency. The other experiment was aimed at verifying if the dyes, pumped with the light from a ruby laser, were able to give the laser effect by themselves.

Sorokin decided to start with the first experiment, sending light from a ruby laser onto the sample. When the spectrum of the emitted light was examined, it was evident the second experiment had succeeded. Placing the dye sample between two mirrors, Sorokin and Lankard obtained a powerful laser beam at 7555 Å. They examined other dyes and discovered the effect was quite general. They tried with all the dyes they could find in the store. One afternoon Sorokin went down the aisle at his lab asking colleagues 'What colour do you want?', so many were the wavelengths that could be obtained by changing the dye. One thing they did not notice was that the new laser could be tuned, that is it emitted at a wavelength which could be varied over a reasonable range using the same material.

In their studies they were preceded, in 1961, by the theoretical considerations of two Russians, S G Rautian and I I Sobel'man, and also by experiments made by D L Stokman with co-workers in 1964 during which some indications were obtained of a possible laser effect with the aromatic molecule perylene pumped with light from a flash lamp.

A little later, and independently, the chemist Fritz P Schafer, then at Marburg University in Germany, while studying the saturation characteristics of certain organic dyes of the cyanine family obtained the same effect. He was studying the light emitted by a dye pumped with powerful Q-switched ruby laser with the help of a spectroscope, when his student Volze, by examining highly concentrated solutions, obtained signals thousands of times stronger than expected. Soon the two researchers understood they were witnessing laser emission, and together with Schmidt, then a doctorate student, they photographed the spectra at different concentrations obtaining the first proof that a laser had been built which was tuneable in wavelength over 600 Å by changing the concentration of the dye or the reflectivity of the mirrors of the resonant cavity. The effect was soon confirmed and extended to a dozen different dyes of the cyanine family. A surge of interest resulted, as happens in these cases, and worldwide research soon identified thousands of dyes showing the laser effect. Finally in 1969, B B Snavely and Schafer obtained an indication of a possible continuous emission, which was obtained the following year at the Eastman-Kodak Research Laboratory by O G Peterson in collaboration with other researchers who used an argon laser to pump a solution of rhodamine 6G dissolved in water.

These lasers allowed a long pursued dream to be realized: to obtain a laser that was easily tuneable over a wide range of frequencies. The dyes are also interesting for other reasons. They can be used dissolved in solids or liquids with easily controllable concentrations. The liquid solutions are particularly convenient. Cooling is obtained by recirculating the solution. Moreover the liquid is not damaged and the cost of obtaining the active material is very low, in contrast to solid state lasers for example. By choosing a suitable dye among thousands available and the wavelength of the pump radiation, it is possible to cover all the visible spectrum up to the near-infrared. A laser tuneable to a desired wavelength was finally born! In the middle of 1967, Bernard H Soffer and Brill B McFarlane, by substituting one of the mirrors of the cavity with a diffraction grating, successfully built a dye laser which could be tuned over a range of more than 400 Å, simply by tuning the position of the grating. These lasers, due to their broad band emission, can be operated in a pulsed regime which may be shorter than a picosecond.

Laser diodes

Semiconductor or diode lasers are important for many applications. They exploit the energetic situation of the delocalized electrons rather than the levels of a particular impurity atom. Like in all solids, the energy levels of electrons are grouped into bands. At the temperature of absolute zero, in semiconductors, all the available levels are filled in one band (valence band) and the successive free levels are grouped into another band (conduction band) which is completely empty, separated from the valence band by a certain energy gap in which there are no available states. The interval between the two bands is called the forbidden energy gap, or simply gap. In these conditions, the material cannot conduct an electric current and therefore behaves as an insulator. As the temperature increases, if the conduction band is separated from the valence band by not too great an energy, thermal agitation is sufficient to cause some electrons to jump into the conduction band. Finding nearly all the levels empty, these electrons are able to carry a current; however, being small in number they give rise to only a small current. Accordingly, the material becomes a poor conductor: a semiconductor. The electrons in the conduction band which are able to carry a current leave vacant states in the valence band. These vacant states, referred to as holes, behave as positive particles and also contribute to the electrical conductivity. In a pure semiconductor, thermal agitation produces electrons in the conduction band and holes in the valence band in an equal number.

The electrons and holes capable of carrying a current are called charge carriers. If for some reason in the conduction band more electrons are present than allowed by Maxwell–Boltzmann statistics, the excess electrons drop into the vacant energy levels in the valence band. In this way an electron returns to the valence band and a hole disappears. The same occurs if, instead, more holes are present in the valence band than those allowed to remain at a given temperature. This process is called recombination of the two charge carriers. It occurs by yielding the energy corresponding to the energy gap between the two levels to the lattice, in the form of mechanical vibrations, or by emitting a photon. In the latter case, the transition is called radiative and the emitted photon has an energy corresponding to the energy difference between the two levels which is roughly speaking equal to the band gap.

Sometimes the semiconductor is not perfectly pure. Impurities create energy levels for electrons that are within the energy gap. If these extra levels are near to the bottom of the conduction band, thermal agitation makes their electrons jump into the conduction band that now can carry an electric current. The impurity levels remain empty and, because they are fixed in the material, they are not available to carry a current. In this way the only current carriers are electrons in the conduction band, and the semiconductor is said to be doped n-type ('n' to remember conduction occurs through the movement of negative charges). Conversely, if the impurity levels are near the top of the valence band, thermal agitation makes electrons from the valence band jump into the impurity levels and the created holes may now carry a current. One speaks then of a p-type semiconductor ('p' for positive charges). It is possible to dope a semiconductor so that there are both p-type and n-type regions, with a very thin interface between them. The interface between the different regions is called a pn junction. If one makes current flow through a junction of this kind, making the n region negative and the p region positive, electrons are injected into the junction. By using this property, at the end of the 1940s transistors were invented which revolutionized the electronics world.

Although semiconductors have been known for some time, their physics was fully understood only after the invention of the transistor in 1948. One can thus understand how in the 1950s there was some hesitancy in their use for making a laser. Semiconductors were in any case the first to be considered as a possible medium in which to obtain radiation by means of stimulated emission. A number of different proposals were put forward at the same time. In 1954, John von Neumann had discussed with John Bardeen (one of the inventors of the transistor) the possibility of using semiconductors. Three years later, in 1957, a true explosion occurred. In Japan, a patent was filed on 22 April 1957 and published later on 20 September 1960 by Y Watanabe and J Nishizawa in which recombination radiation produced by injection of free carriers in a semiconductor was considered. The patent title was 'semiconductor maser' and, as a specific example, recombination radiation at about 4 µm, i.e. in the near infrared, in tellurium was considered. Naively, the authors considered the semiconductor in a resonant cavity of the kind used in the microwave region, but the concept of using an injection of carriers and their recombination radiation was sound. At the Lincoln Laboratory at MIT, the physicist Benjamin Lax, inspired by a seminar delivered at MIT in 1957 by Pierre Aigrain (1924–2002), the future Undersecretary of State for research in France, then at Ecole Normale Supérieure in Paris, began to explore transitions in a group of energy levels created when a semiconductor is subjected to a strong magnetic field, in a similar way to the operation of Bloembergen's three-level maser. Aigrain's ideas were presented at an international conference on solid state physics in electronics and telecommunications, held in Brussels in 1958, where he discussed the possibility of using semiconductors to extend maser action to the field of optical frequencies to obtain certain advantages. However, the paper was not published in the proceedings of the conference.

In the ex-Soviet Union a group at the Lebedev Institute of Physics (FIAN) of the Academy of Sciences of the USSR, headed by N G Basov, with B M Vul and Yu M Popov, started in 1957 to consider the possibility of using semiconductors to extend maser emission towards the optical region. Basov began considering the problem together with Popov, who was then working in the luminescence laboratory. The two researchers had first met as students at the Institute of Physical Engineering. Semiconductor physics was studied at FIAN in Vul's laboratory, and he was therefore naturally associated with the research. The collaboration of the three scientists resulted in a proposal of a laser system utilizing an electric discharge in a semiconductor. The proposal was published in June 1958 and was discussed by Basov in the West at the first conference on quantum electronics organized by Townes, in 1959, in the United States. The paper was not in the programme and was presented during a dinner (a semiconductor laser working along these lines was built many years later, in 1968, by Basov). Later, in 1960–61, the group proposed three further methods of excitation: pumping with an electron beam, optical pumping and pumping through electron injection through a pn junction. The authors of these proposals were Basov, O N Krokhin and Popov. Some experimental measurements were also performed. The previous year, in 1959, still at FIAN, under the direction of Basov, a programme called 'Photon' had begun which represented the first Soviet scientific programme for the development of the laser.

The possibility of using semiconductors was further considered in America and was discussed at MIT in 1959 by two other American researchers, N Kromer and H J Zeiger. In 960, William S Boyle and David G Thomas at Bell Laboratoriess filed a patent for the use of semiconductors to build a laser.

An important theoretical result was eventually obtained in 1961 by two French researchers, Maurice Bernard and G Duraffourg at the Centre National d'Etudes des Télécommunications (CNET) at Issy-les-Moulineaux. They presented a complete and exhaustive theoretical discussion from which the possibility of stimulated emission in semiconductors via transitions between conduction and valence bands was demonstrated, obtaining the fundamental relationship which must be fulfilled to produce a laser effect. These authors also considered certain materials in which this condition was likely to be fulfilled and suggested, among other compounds, the semiconductors GaAs (gallium arsenide) and GaSb (gallium antimonide). After the publication of this work, many groups started active research. In January 1962, the Russian D N Nasledov, along with some colleagues from the A F Ioffe Physico-Technical Institute in Leningrad, reported that the linewidth of the radiation emitted from GaAs diodes at 77 K narrowed slightly at high current densities. They suggested that this might be a sign of stimulated emission. In the United States, some groups had started experimental work more or less at the same time with their efforts also focused on developing the first semiconductor laser device. The groups were IBM, RCA, the Lincoln Laboratory at MIT and General Electric in two different laboratories in Schenectady and Syracuse. The competition to be first to obtain laser action became a frantic race, briefly described here.

At the Watson Research Center of IBM, Rolf W Landauer (1927–1999) in December 1961 formed a small group to examine the problem in a systematic way. William P Dumke, also from IBM even if not a direct collaborator of Landauer, showed that simple semiconductors like silicon and germanium, which are the most commonly used for electronics applications, would not be suitable, due to their band configuration, and suggested the use of more complex semiconductors from the structural point of view, like gallium arsenide, in which the minimum energy of the conduction band just coincides with the maximum of the valence band (the so-called direct gap semiconductors). IBM was well placed to consider gallium arsenide because it had already started a programme to study it for electronics applications.

Compound semiconductors and especially gallium arsenide were also being studied at General Telephone and Electronics Laboratories (GT&E) at Bayside, New York. Here Summer Mayburg headed a small group which studied which kinds of devices could be made with gallium arsenide. In March 1962, at an American Physical Society meeting, Mayburg presented a paper on the electroluminescence of GaAs diodes at 77 K, that is on the radiation that these diodes, cooled at the temperature of the liquid nitrogen, emitted when they drew an electric current, and on the invitation of Landauer, who had first met in their college days, visited the IBM laboratory. The work presented by Mayburg reported results obtained by Harry Lockwood and San-Mei Ku, who had shown that under certain circumstances practically every charge injected through a pn junction resulted in a photon. This result was analogous to that found by Maiman in ruby and indicated that pn junctions were the ideal system in which to obtain laser action.

Jacques I Pankove at RCA had spent one year in Paris during 1956–57, working with Aigrain. Coming back from France he had started research but without much support because the Head, William Webster, considered that the commercial income from semiconductor laser sales would be very low. In January 1962, at the American Physical Society conference, Pankove announced that, together with his colleague M J Massoulie, he had obtained recombination radiation from gallium arsenide junctions. Mayburg acknowledged he could be beaten by RCA and redoubled his efforts.

At IBM, after the Mayburg seminar, the theoretician Gordon Lasher started to study how to make a cavity for a semiconductor laser and at the same time in the nearby IBM laboratory in Yorktown Heights in New York state, Marshall I Nathan started thinking how to build a gallium arsenide laser.

At the MIT Lincoln laboratory, a group headed by Robert H Rediker had been studying gallium arsenide diodes since 1958. In mid-1961 they started to measure recombination radiation as a means of investigating diode characteristics and found the same strong luminescence already observed by GE and RCA.

In July 1962 Mayburg's results were discussed at a Solid State Device Research Conference at the University of New Hampshire, and R J Keyes and T M Quist of the Lincoln Laboratory announced they had built gallium arsenide diodes with a luminescence efficiency that they estimated to be 85%. Pankove, in May, had presented similar results at another conference. MIT used the luminescence emitted by a diode to transmit a television channel, and the news was reported in the *New York Times*.

At this point a fourth group entered the race. Robert N Hall from General Electric at Schenectady had attended the 1962 conference in New Hampshire and was struck by the presented results. The high efficiency with which light was emitted by gallium arsenide pn junctions impressed him greatly, and returning by train he started to make calculations of the possibility of building a laser and thought about how to obtain a Fabry–Pérot cavity. The idea was to make a pn junction, cut it and polish it. Hall had been an amateur astronomer and at school had built a telescope; he knew how to polish optical components. Nowadays cavities are made by cleaving the crystal in the right directions, but at that time he was not familiar with such techniques. After having approached some of his colleagues, he obtained permission from his boss to start work on this project. The principal difficulty was to make a junction in GaAs which, to satisfy the Bernard and Duraffour conditions, needed to be heavily doped. A second difficulty was to cut and polish the structure and make the cut sides highly parallel to each other. It was then necessary to send a very high current through the junction to inject a sufficiently high number of electrons, and this current had to be pulsed for a very short time to avoid melting the structure. To prevent the temperature increasing too much, it was necessary to cool to liquid nitrogen temperature (77 K).

Even though Hall had been the last to enter the competition, he was the first, if only by a short while, to obtain in September 1962 the first laser diode. Bernard visited Hall's laboratory several times, discussing the possibility of semiconductor lasers. During one visit he happened to appear just after Hall's group had got the first one going, but before they had submitted the paper describing it. The achievement was therefore kept secret to avoid being beaten. Hall had to discuss the problems of how to make a laser without being able to tell Bernard that he had one working in the next room.

The New Hampshire conference also inspired N Holonyak, an expert of gallium arsenide in General Electric's Syracuse laboratory. When the first diode was operated, nearly simultaneously several groups announced laser action in pn junctions in GaAs. In all cases the GaAs was cooled at 77 K and pumped with current pulses of high intensity and short duration (a few microseconds). The laser realized in the General Electric laboratory at Schenectady was announced in a paper received on 24 September 1962; a second laser was announced on 4 October by the group with Marshall I Nathan of IBM at Yorktown Heights and a third one on 23 October by a group at MIT's Lincoln laboratory, headed by T M Quist. At General Electric laboratories in Syracuse, Holonyak had realized his own laser, submitting his paper on 17 October. All these lasers were made with a gallium arsenide junction cooled at liquid nitrogen temperature and pumped with current pulses of high intensity and duration of a few microseconds.

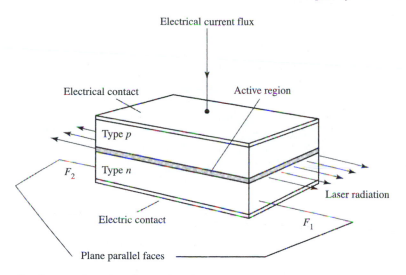

Figure 60. Layout of a pn junction semiconductor laser of the simplest type. Laser radiation is emitted in the thin active region between the p and n zones and is reflected back and forth by the two plane parallel facets F_1, F_2 which act as mirrors.

Hall's device was a cube of 0.4 mm edge with the junction lying in a horizontal plane through the centre. The front and back faces were polished parallel to each other and perpendicular to the plane of the junction creating a cavity. In many studies, it had been speculated that the mirrors would be parallel to the junction plane, so the radiation would be emitted perpendicularly to the junction. Hall's laser, as well as all the other working lasers, comprised an arrangement of the Fabry–Pérot mirrors so that the radiation would bounce back and forth in the junction plane, that is the region where the injected carriers recombine and emit light. This gave a relatively long path for amplification (figure 60).

The laser was driven with current pulses of 5–20 ms duration, obtained by connecting the positive pole to the p-doped side and the negative pole to the n-doped side with the diode immersed in liquid nitrogen. When the current reached the very high value of 8500 amperes per square centimetre (A/cm^2), the laser action started as could be seen by a sudden step increase in the emitted radiation which also showed a line narrowing from 125 to 15 Å.

Nathan used a somewhat different system employing a junction without resonant cavity. The threshold for laser action, still at liquid nitrogen temperature, was obviously higher at between 10 000 and 100 000 A/cm^2. Quist used a structure of $1.4 \times 0.6\,mm^2$ with the short sides being polished optically flat and nearly parallel. At liquid nitrogen temperature, the threshold was about 1 000 A/cm^2. Finally, Holonyak and Bevacqua used a junction in a compound of gallium arsenide and phosphide. The diodes

257

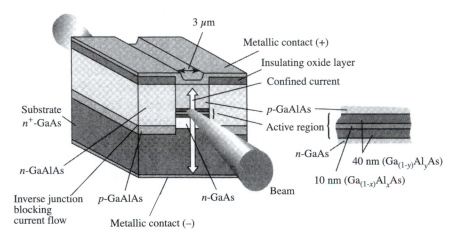

Figure 61. Buried heterostructure laser diode. (From *Laser Focus World*.)

were rectangular parallelepipeds, or cubes, with two opposite, parallel sides carefully polished so as to give a resonant cavity. Using this material they were able to obtain emission in the region 6000–7000 Å instead of about 8400 Å as obtained by all the other diodes using simple GaAs.

In Russia, a short while after the construction of the American lasers, V S Bagaev, N G Basov, B M Vul, B D Kopylovskii, O N Krokhin, Yu M Popov, A P Shotov and others, built a laser diode at FIAN. The results were discussed at the third conference on Quantum Electronics in Paris in 1963.

The first lasers made with one material for the n and p sides of the junction had high thresholds. In 1963 Kroemer suggested using hetero-junctions, in which a layer of semiconductor with a relatively narrow energy gap is sandwiched between two layers of a wider energy-gap semi-conductor. A similar suggestion was made at the same time by Zh I Alferov and Kazarinov at the Ioffe Physico-Technical Institute in Leningrad. The two Russian researchers did not publish it. Six years elapsed until Hayashi and Panish of Bell Laboratories and Kressel and Nelson at RCA developed the first heterostructure lasers. At the same time Alferov and collaborators developed the more complex multilayer structures which are nowadays called double-heterostructure lasers. These efforts were awarded with the physics Nobel prize. In 2000, the prize was awarded to Zhores Alferov (1930–2002) and Herbert Kroemer (1928–) both 'for developing semiconductor heterostructures used in high-speed and opto-electronics' and Jack Kilby (1923–) 'for his part in the invention of the integrated circuit'.

Alferov was born in Vitebsk, Byelorussia, and graduated in 1952 from the V I Ulyanov Electrical Engineering Institute in Leningrad entering the

Physico-Technical Institute in 1953. He was director of the Institute from 1987 and a member of the Russian Duma.

Kroemer was born in Weimar, Germany in 1928 and earned his PhD from the University of Göttingen in 1952 with a dissertation on the then new transistor. In 1968 he joined the faculty of the University of Colorado and in 1976 moved to the University of California, Santa Barbara.

The development of semiconductor lasers has been slow owing to various reasons. It was in fact first necessary to develop a new technology for manipulating semiconductor materials, considering that silicon, for which there already existed a well-established technology, could not be used. There were also problems connected with the high currents needed to obtain laser action that limited the operation to short pulses and required the use of very low temperatures, resulting in a low efficiency of the device. A notable step forward for the solution of these problems was made in 1969 with the introduction of heterostructures. In a heterostructure laser the simple pn junction is replaced by multiple layers of semiconductors of different composition. The active region is then reduced in thickness and the total current required for laser operation decreases notably, with a corresponding decrease of the heating effect. It followed that it was no longer necessary to cool and one could achieve a working laser even at room temperature (figure 61).

Two factors are largely responsible for the transformation of semiconductor lasers from laboratory devices working only at very low temperatures to practical opto-electronic devices able to work continuously at room temperature. One is the exceptional and lucky similarity of the lattice constants of aluminium arsenide (AlAs) and gallium arsenide (GaAs) that allows heterostructures to be made by layers of different composition of the compound gallium and aluminium arsenide ($Al_xGa_{1-x}As$). The second is the presence of many important opto-electronic applications for which semiconductor lasers are particularly suited given their properties: small dimensions (a few cubic millimetres), high efficiency (sometimes more than 50%), pumping obtained directly from the current and longer life compared with other lasers. The fact that the laser is pumped directly by the current allows direct modulation of its output, simply by modulating the current. This characteristic is ideal for information transmission systems.

Does the laser exist in nature?
The answer seems to be yes! A laser emission at a wavelength of about 10 µm (a typical emission line of carbon dioxide) was identified in the Mars and Venus atmospheres in 1981 by Michael J Mumma of the Laboratory for Extraterrestrial Physics of the NASA Goddard Space Flight Center. The emission had been already observed by some of Townes' students in 1976, but was confirmed to originate from a natural laser in 1981.

The population inversion of carbon dioxide that forms the great part of the atmosphere of these planets is produced by sunlight and therefore occurs only in the illuminated hemisphere. The mechanism is exactly similar to that of carbon dioxide lasers that are built on Earth and emit at $10\,\mu m$ and are used as power lasers for mechanical cutting and other applications. The emission lines in the atmosphere of the two planets are about 100 million times more intense than could be obtained if the gas was in thermodynamic equilibrium conditions at atmospheric temperature. A part of the observed light is amplified radiation by the inverted medium. If two mirrors could be placed in orbit around the planets, oscillations could be obtained just as on Earth.

The technology to realize lasers on a planetary scale is out of our grasp at present, but what the future may bring we do not know. The emission lines have been useful to measure the temperature and the winds on Mars and Venus.

Masers in space have been found for many years, as we saw, and there is no reason why one could exclude the existence of lasers. The process, however, because it requires greater photon energies, is relatively more difficult. Early in 1995, a team of astronomers detected amplified infrared light coming from a hydrogen disc whirling around a young star of the Cygnus constellation, 4000 light years away from us. The intensity of the emission at one wavelength compared with its neighbours shows that one is in the presence of stimulated emission (figure 62). Previous observations of the

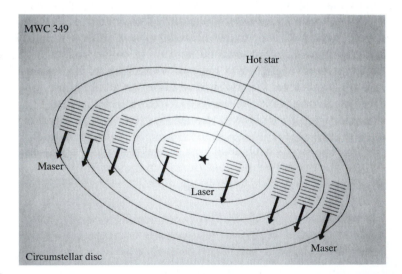

Figure 62. The natural laser in the star MWC349. The laser emission occurs in the regions of the hydrogen disc that are nearer to the central star, while the maser emission occurs in the far regions. The radiation is emitted in the figure plane and reaches Earth which by chance lies in the same plane.

star, called MWC349, had already revealed in 1994 an intense maser emission from its disc, at wavelengths of 850 and 450 μm, produced by hydrogen. The study of the processes according to which this emission occurred suggested that also emission at shorter wavelengths could be present in disc regions nearer to the star. Therefore Vladimir Strelnitski of the National Air Space Museum of Washington, Edwin Ericson and Michael Haas of the Ames Research Center of NASA and Howard Smith and Sean Colgan of the SETI Institute of Mountain View in California, placed an infrared telescope on an aeroplane, let it fly at a height of 12 500 m to avoid the absorption of infrared radiation by our atmosphere, and observed a line at 169 μm that is six times more intense than what one would expect by an emission in thermal equilibrium. This line is produced by the hydrogen atoms that have been ionized by intense ultraviolet light coming from the star or through more complex processes that occur in the disc. When the free electrons recombine with the ions they emit photons. Most of the emission occurs spontaneously, but it is possible also to have stimulated emission. The same process produces maser emission in other parts of the disc, but in the central parts one observes laser emission, partly because the hydrogen there is more dense and partly because the ultraviolet pumping light is more intense. By chance the disk is oriented almost edge-on so that some of the light aims at the Earth and therefore the emitted laser beam can be received. The disk is a region where one believes that planets could form and the observed emission comes from one part of this 'nursery of planets' that is at the same distance from the star as the distance between the Earth and the Sun. Laser emission may therefore help our understanding of the state of the gases in the disc. The wavelength of 169 μm, in the medium infrared, is in a region that is really between sub-millimetre waves and the region typically assigned to optics. One may thus speak of a laser at huge wavelength or of a maser at short wavelength, according to how the situation is interpreted.

Lasers in the ultraviolet region exist too. An emission in this region has been identified using the Hubble space telescope, from a gaseous cloud near to the star η-Carinae.

We may therefore conclude also that celestial lasers, like masers, are objects already existing in the cosmos and therefore it would be more exact to say that masers and lasers have not been invented but discovered.

CHAPTER 14

A SOLUTION IN SEARCH OF A PROBLEM OR MANY PROBLEMS WITH THE SAME SOLUTION? APPLICATIONS OF LASERS

In 1898, in his book *The War of the Worlds*, H G Wells deemed the Earth invaded by Martians who were using a death-ray which was able to pass without any difficulty through bricks, set fire to forests and burn steel as if it was paper. Weapons of this kind had been used by certain cartoon characters before and soon after the Second World War, and inspired the military to dream about future weapon systems. Nowadays very powerful laser beams can achieve similar results in reality.

Once the first laser was built, cartoon writers used it instead of the obsolete ray-guns of their characters and the tabloids indulged in savage hypotheses on the possibilities of developing laser cannons and other farfetched weapons. Arthur Shawlow demonstrated the possibilities offered by the ruby laser at conferences by using a gun with a small ruby laser inside to explode a small blue balloon, and went on to assemble a collage of media fantasies and glued it to his office door at the University of Stanford with the wording 'The incredible laser' underneath which he wrote 'For the credible lasers, look inside'.

The light of a laser differs from that of an ordinary light source more or less as sound differs from noise. Moreover a laser beam may propagate for kilometres, increasing its diameter only by a small amount. For example, when in 1969 a ruby laser beam was sent to the Moon to be reflected back towards the earth by a special mirror positioned by Armstrong, the spot on the Moon had a diameter of only 9 km.

Another quality of lasers is their huge luminous intensity or brightness. While we warm ourselves in the midday sun on a beautiful summer's day, on our finger falls sunlight whose power is approximately a tenth of a watt. The light from a laser may be concentrated to a point on our finger which would receive 10^9 W!

These properties and the huge number of available lasers have resulted in a number of applications in very different fields with many surely still to come. This is very different from the position immediately after the invention of the laser when no application was yet known and people were saying that the laser was the bright solution of a problem which did not exist (a solution looking for a problem).

The variety of different kinds of lasers range from those as big as a football field to others as small as a pin-head. The light they emit ranges from the invisible infrared to the invisible ultraviolet, down to x-rays through all the colours of the rainbow. Some are wavelength tuneable and their intensity may extend for several orders of magnitude.

Some emit pulses as short as a femtosecond (10^{-15} s) and others may emit continuous beams that last for decades. Like Wells' rays, some lasers may focus their light to a point bright enough to vaporize steel or any other material by concentrating energy on its surface a million times more rapidly and intensively than a nuclear explosion. Others emit energy that is not even sufficient to cook an egg.

Few scientific developments have excited the imagination of scientists and engineers as much as the laser. The laser has made it possible to investigate experimentally optically excited plasmas, with the hope, not yet realized, of obtaining controlled nuclear fusion. It has made possible the generation of optical harmonics, that is new laser wavelengths, photon echoes, self induced transparency, ultra-short optical pulses, optical shocks, self trapping of optical radiation, optical parametric amplification, measurements on the Moon, very high resolution spectroscopy, highly refined measurements of many physical constants, studies of ultra-fast processes in atoms, molecules, liquids, solids and gases. It is possible to use lasers to print journals and magazines. Using lasers the police may detect fingerprints left even 40 years previously. Holograms allow three-dimensional images to be created. Projectors that use lasers emitting different colours are used in presentation graphics and for the entertainment industry. Laser light is also used for special effects in movies and video, in rock concerts, in night-clubs and in commercial shows to create lighting effects.

At the very beginning there were very few possibilities for the commercial exploitation of lasers, except selling their components to researchers involved in laser construction and in their development. Companies that wanted to work in the field had to develop their own applied research in order to obtain contracts, essentially from the military. The sentence 'the laser is a solution in search of a problem' was common at that time. Then laser technology entered a development phase in which applications were sought. Many companies during this phase could not or decided not to pursue a technology that still had no definite clear applications and withdrew from the field. Many entrepreneurs, on the other hand, redoubled their efforts and tried to find applied and commercial avenues. Nowadays laser technology has broken through. Many applications can be identified in which the laser boasts lower cost or better capacity than older technologies. The laser has today found markets in the field of telecommunications, in the treatment of information and in its storage, in entertainment, in print, in material processing and in medical applications, to mention just a few. More markets will arrive in the future. The layman profits by laser technology in technologies such as video and audio discs, laser printers for computers, bar code readers for prices in shops, optical fibres for telecommunications and in several medical treatments.

New words describe the use of light in its various applications. Electronics is the name commonly employed to describe the behaviour and applications of electrons. The term was already in use in 1910. When the maser was discovered, that is an electronic device for which knowledge of quantum mechanics is necessary, the term quantum electronics was coined. This term was then extended to all electronics devices whose understanding requires quantum mechanics, as for example transistors. Opto-electronics is a term of more recent coinage (it was used for the first time in 1955 even before the invention of the laser) and describes the phenomena and those devices whose operation occurs through the combined use of electronics and optics. Many of the devices that nowadays use the laser are typical opto-electronic devices and the same laser may be described as an opto-electronic device. To describe more specifically the application of photons to devices, especially in the field of transmission of information, one has used since 1952 the term photonics meaning, in analogy with the term electronics, the application or the production of photons for the transmission of information and phenomena such as the production of photon beams (light), the guided propagation of light, its deflection, modulation and amplification, the optical treatment of images, the storage and detection of light signals. As one may note there are no clear boundaries between one term and another and often these different terms are used in an interchangeable way. With time one may arrive at more precise definitions of each term.

In 1984 the global market for lasers was more than two million euros for the commercial sector and an additional one million for the military sector. In 1994 the total market was 1 billion euros for laser retail only. During this escalation of success and applications, comical incidents were not absent. For example, in the 1970s an employee of American customs decided that lasers were harmless and could be imported or exported without limitation, but not the laser beams themselves!

At this point we wish to mention some of the huge possibilities of lasers by describing some of the applications of greatest interest both from an historic and modern standpoint.

The laser and the military
Even before lasers were actually built, they had succeeded in attracting some interest from the military for a long series of applications that might seem feasible. One recognized that the high directionality of a laser beam maintained the secrecy of a signal sent by modulating for example its intensity, because unless one lay directly along its path it would not be possible to

264

detect it laterally. Additionally, the possibility of focusing the beam on the target avoids all losses of power that one suffers if the beam is emitted in all directions, as happens with radio waves and also, to a lesser extent, with microwaves. It seemed then that the laser could provide a unique means for optical communications or even for transmitting energy. However, the first experiments, performed as soon as lasers were available, showed that the Earth's atmosphere has detrimental effects on the propagation of the beam: the light is absorbed or scattered by atmospheric constituents; if it is raining or snowing or if there is a strong fog, it is unable to pass, but even if the atmosphere appears clear, its properties are degraded. For example, its intensity does not remain constant but begins to fluctuate casually with time due to a phenomenon known as atmospheric turbulence. This is a phenomenon well known to astronomers who are used to observing stars whose images fluctuate in time (they call this effect scintillation). However, if laser propagation in the Earth's atmosphere was revealed to be disastrous, this was not so for its propagation in vacuum, such as between satellite and satellite or for example a satellite and the Moon, or even for short propagation distances such as in airports. An excellent way to propagate light signals without appreciable losses was obtained by employing optical fibres, in which the electric signals transmitted through copper wires or with radio waves are substituted by light signals propagated in special glass fibres capable of transmitting information from one continent to the other. Optical fibres have in fact better characteristics than copper wires, weigh less and cost less.

Military personnel also had in mind other applications, for example radar. A radar with an optical carrier allows a precision and an ability to 'see' details of the target unimaginable even with millimetre waves. It is possible also to measure the velocity of a target. On the other hand, changes produced by the deleterious effects of the atmosphere on optical radar beams may be used to measure the properties of the atmosphere itself (lidar), such as ozone concentration, pollutant build-up, etc. or atmospheric turbulence that is very important information for aircraft navigation.

Range-finding was the first military application of the new laser technology. Operational range finders were introduced into the armed forces as early as the mid-1960s. The principle of operation of these systems is very simple. A short laser pulse—about 10–30 ns—is emitted towards the target, and the time is measured until a reflected signal is returned. Since the pulse travels with the speed of light, the distance is immediately derived. Using a sequence of pulses, laser trackers can be built.

Lasers may guide weapons. In 1965 the specialized press reported that the American Forces were investigating a hand-held laser that could be used to guide a shell onto the target. Soon after, experiments were described in which the laser was used to illuminate small targets and provided accurate guidance to supersonic missiles. The first target laser designators were used in 1972, near the end of the Vietnam War. 'Intelligent' bombs guided by lasers heralded the advent of weapons guided with very high precision. It was a turning point for military technologies because the new guiding systems improved the probability that each intelligent bomb would strike the target compared with 'stupid' bombs launched by an aircraft. In the Persian Gulf War and in the Serbian-Bosnian War, these laser-guided weapons were the norm. In 1968 the US Navy began researching the use of high power lasers as weapons, succeeding in shooting down its first target missile in 1978. Then the Armed Forces investigated the possibilities of using lasers to disable sensors and even to blind enemy soldiers. The Reagan administration launched a programme of antimissile defence based on the use of lasers, known by the name of 'Strategic Defence Initiative', presented in the famous 'Star Wars' speech given by Reagan on 23 March 1983. The programme that should have provided laser systems capable of detecting enemy missile attacks and destroying ballistic missiles as they flew encountered several criticisms and, in the end, it was little developed. The Clinton administration modified the defence strategy by founding the Ballistic Missile Defence Organization, with less ambitious purposes but better probability of success.

Most defence systems are designed to destroy a missile warhead before it can reach the target. The systems already developed intercept the missile during the terminal phase of its path, after it has re-entered the atmosphere. Other systems will try to intercept the missile outside the atmosphere or even in the phase that immediately follows its launch.

Once the missile has been fired the following defence action take place. First infrared sensors on geosynchronous orbit satellites detect the tail of hot gases exiting the missile while it climbs beneath the clouds. The satellite alerts military commands that a launch has been exacted and indicates the general area towards which the missile is directed. This information is used to point the defence sensors towards the right spot for tracking. The sensors track the target, identify the armed head and pass on this information to the interceptor. Traditionally these sensors are Earth-bound radar but in the future they will be helped by satellites in very low orbits equipped with infrared sensors. On the basis of the information, the interceptor flies towards the point where the intercept has been calculated receiving up-to-date data on the coordinates during its flight and detach the destroyer that uses its own guidance systems to centre the target.

This complex dance that we have simply sketched here must be choreographed by an extremely sophisticated battle management system with very fast response times. The total flight time for a Scud with a range of 300 km is at most 4 min. Missiles with a longer range stay in flight for 15 min or less.

One of the possible scenarios is one in which the missile is destroyed during the launch phase. One of the advantages of this is that missile engines emit hot gases which are easy to identify. However, one needs to be sufficiently near to the missile to intercept it, because engines burn at most for a few minutes. In that short time period, the defence system must individuate the launch, follow it for a sufficiently long time to calculate its trajectory and eventually intercept it.

The Pentagon has also developed a system to achieve this: the Air Force's Airborne Laser. It is a chemical laser, borne by a 747 Boeing plane that should intercept missiles at a distance of several hundred kilometres. Interest in using a laser beam centres on the fact that it travels at the speed of light so saving precious fractions of a second. The laser attacks missiles by focusing its beam on their metal surface, heating it until the strength of the metal falls dramatically. Each kind of metal has its own characteristic failure point: 460 °C for steel and 182 °C for aluminium. There are at least two possibilities. The first, for missiles carrying liquid fuel, is to heat the fuel tank to the point of structural failure, whereupon it will rupture from its internal pressure. The other is to heat a sufficiently large arc along the missile circumference to the critical structural-failure temperature. When this is reached, the aerodynamic and inertial forces acting along the missile's axis will bend the structure in half (figure 63). On board the plane will be three major laser systems. The first, that produces the killing beam, is a continuous laser. There are then two pulsed lasers, one to track and keep a lock on the target, and another to minutely adjust the properties of the primary beam in response to atmospheric changes (see the section on adaptive optics, p 281). The heart of the system is clearly the primary killing laser. This is a chemical oxygen iodine laser positioned in the rear of the plane and

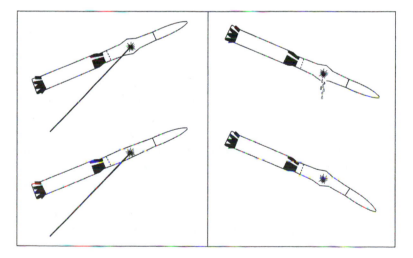

Figure 63. The left part of the figure shows the laser beam which warms the liquid propellant tank (top) or an arc on the side of the missile (bottom). As a consequence (on the right) the tank breaks and missile falls (top) or it collapses on itself (bottom).

producing a continuous infrared beam with a wavelength of 1.315 µm with power of the order of several megawatts, considerably higher than has been produced by similar lasers up to now. In the chemical oxygen iodine laser, excited iodine atoms are produced in mutual collisions with excited oxygen molecules obtained in a chemical reaction. The system was under study in 1997 and has received severe criticism.

Solar lasers

Immediately after the discovery of the laser one started to dream of converting directly white incoherent sunlight into the monochromatic coherent light of a laser. This would allow, for example, enormous reduction in the weight of the laser on a satellite, because the function of all the peripheral pumping machinery would be provided by the Sun. These dreams were soon realized and in 1966 lasers pumped by the sun were built. However, their efficiency was rather low, typically of the order of 1%, that is only one hundredth of the collected sunlight was transformed into laser light. Since then very sophisticated optical methods of concentrating sunlight to a previously unimaginable extent have been developed. One result was a concentration of 72 W per square millimetre, that is greater than the intensity of sunlight at the Sun's surface (63 W/mm^2). With these intensity values of pump light it is possible to make lasers with improved efficiency and one may expect reasonably good results. Efficiency has already been raised to more than 6%.

Optical fibres and laser telecommunication

Since antiquity light has been used to send messages. China, Egypt and Greece used smoke by day and fire by night to transmit signals. Among the first historical records of optical communication we may recount the siege of Troy. In his tragedy *Agamemnon*, Aeschylus provides a detailed account of a chain of fire signalling positions on the summits of Mounts Ida, Anthos, Macisto, Egiplanto and Aracnea and the cliffs at Ermea of Lemno and Cithaeron to transmit to Argos the news of the Troy capture by Acheons.

More recently, even though still in ancient times, the Emperor Tiberius, whilst in Capri, used to communicate with the mainland via light signals. In Capri the ruins of the ancient 'Faro' (light) can still be seen near the Tiberio emperor's villa on Tiberio Mount.

One of the first optical communication systems in North America was established about 300 years ago in the colony of New France (today Quebec). The regional government, afraid of the possibility of an English naval attack, established a number of positions for signal fires in many villages along the Saint Lawrence River. There were no less than 13 sites in the chain that began in Ile Verte about 200 km downriver from Quebec. Beginning in the early

1700s a guard was posted at each village every night during the navigation season to watch for any signal coming from the next village downriver. In such a way, the British offensive of 1759 was communicated to Quebec before it was too late.

In 1790 a French engineer, Claude Chappe, invented semaphores mounted on a series of towers that could send messages from one tower to the other. In 1880, Alexander Graham Bell (1847–1922) obtained a patent for the 'photophone', an instrument that made use of reflected sunlight to transmit a sound to a receiver. The reflected light was intensity modulated by the vibration of a reflecting diaphragm placed at the end of a tube into which Bell was speaking. The modulated light was then transmitted to a detecting mirror positioned at a distance of 200 m and focused onto a suitable receiver. Although Bell considered the photophone his most important invention, its practicality was limited by the fact that sunlight is not always available and the Earth's atmosphere is not always clear. These considerations did not, however, prevent Bell writing to his father:

'I have heard articulate speech produced by sunlight! . . .
Can imagination picture what the future of this invention is to be! . . .
We may talk by light to any visible distance without conducting
wire. . . . In warfare the communication of any army could neither cut
nor tapped'.

The invention of the laser stimulated anew interest in optical communications. However, it was soon demonstrated that the Earth's atmosphere disturbed laser light propagation in an unacceptable way. Various structures involving empty tubes with gaseous lenses and dielectric waveguides were considered during the early 1960s but were all discarded when, at the end of the 1960s, low loss optical fibres were developed.

The realization that thin glass fibres may guide light due to total reflection was an old idea, known in the 19th century by the English physicist John Tyndall (1820–1893), and had been used for many years for the construction of instruments and for illumination. However, in the 1960s even the best glasses exhibited enormous attenuation which caused the transmitted power along the fibre to decrease strongly with distance. At that time a typical value of attenuation was one decibel per metre, meaning that after travelling 1 m the transmitted power was reduced by 80%. In such circumstances only transmission over distances of the order of a few tens of metres was possible and the sole possible application was in the medical field, for example for endoscopes. In 1966 Charles K Kao and George A Hockham, at Standard Telecommunications Laboratory in Great Britain, published a fundamental paper in which the authors observed that using silica glass, if one had carefully controlled the impurities in the glass and if one had covered the fibre with a cladding of a refractive index lower than that of the central fibre, attenuation of the order of 20 dB/km could be achieved, indicating that after propagation for 1 km the beam power had reduced to one hundredth of the input power. Although this is a very low value it was still acceptable for certain applications.

As always happens in such situations, a huge research effort began in Great Britain, Japan and the United States aimed at obtaining fibres that bettered this specification. The first success was reached in 1970, by E P Kapron, Donald Keck and Robert Maurer at Corning Glass Works. They obtained fibres with an attenuation of 20 dB/km at a wavelength of 6328 Å (the wavelength at which a visible helium–neon laser emits in the visible). In the same year, I Hayashi and collaborators announced the first laser diode operating at room temperature.

In 1971, Ira Jacobs was appointed director of the Digital Transmission Laboratory at AT&T Bell Laboratories, at Holmdel in New Jersey, and was charged with developing high transmission capacity systems, and his bosses, Warren Danielson and Rudolph Kompfner, relocated some personnel of his laboratory to the Communication System Research Laboratory, directed by Stewart Miller, in order to keep an eye on what was happening in the field of optical fibres. Three years later, Danielson and Kompfner asked Jacobs to form a study group to examine the practical possibilities of transmission in fibres. It was clear that the most economical initial application for systems utilizing light had been to connect telephonic switchboard centres in metropolitan areas. In the connection cables then in use, the information was transmitted in digital form, that is by coding the signal as a series of pulses, and fibres with their ability to transmit a great quantity of information were the ideal medium for substituting these electric cables. The metropolitan offices were a few kilometres apart and these distances could be covered without problem even by the fibres of the time that had relatively high attenuation. Thus a preliminary experiment was made in Atlanta in mid-1976 with optical fibre cables inserted into the tubes of the ordinary cables in a Bell Laboratories parking area. The entire success of this trial resulted in the installation of a system that connected two telephonic offices in Chicago. On the basis of the results obtained in these first connections, in the autumn of 1977 at Bell Systems, it was decided to develop an optical system for general application. In 1983 a connection was established between Washington and Boston even though there were many difficulties. The system worked at 90 Mbits/s using multimode fibre at a wavelength of 825 nm.

Meanwhile, the Nippon Telegraph and Telephone Company in Japan drew fibres with only 0.5 dB/km loss at 1.3 and 1.5 μm, and the Lincoln Laboratories at MIT demonstrated operation of an InGaAsP laser diode able to emit continuously in the region between 1.0 and 1.7 μm at room temperature. The use of low loss fibres at 1.3 μm made it possible to build better systems. Systems transmitting 400 Mbits/s in Japan and 560 Mbits/s in Europe were built; the European system could transmit 8000 telephonic channels simultaneously. In the United States the companies affiliated to Bell have deposited more than three and a half million kilometres of fibres. The only track that is still in copper is the connection between home and the nearest telephonic office. This 'last mile' as it is named is now being increasingly replaced by fibres. In September 2002, fibres were reaching homes in only 50 communities in the United States. This is a minuscule number compared with the total but indicates at least that the last stage is underway!

The first trans-Atlantic copper cable for telegraphic communications was put into operation in 1858. Nearly one hundred years later, in 1956, the first telephonic trans-Atlantic cable was eventually established and given the name TAT-1. In 1988 the first generation of trans-Atlantic optical fibre cables began operating (they were called TAT-8). They worked at a wavelength of $1.3\,\mu m$ and connected Europe, North America and the East Pacific. In 1991 the positioning of the second generation of fibre cables began, TAT-9, which works at $1.5\,\mu m$ and connects the United States and Canada with Great Britain, France and Spain. Another link is operating between the United States and Canada and Japan.

There are a number of other fibre-optic links in the world. To give an example an optical submarine link between England and Japan covering $27\,300\,km$ in the Atlantic Ocean, the Mediterranean, the Red Sea, the Indian Ocean and the Pacific has $120\,000$ circuits on two pairs of fibres. In comparison the first transoceanic telephonic cable of 1956 allowed 36 conversations and the first optical cable installed across the Atlantic allowed $80\,000$.

Nowadays, after 30 years of research, optical fibres have reached their physical limits. Fibres made with silica can transmit infrared pulses at the wavelength of $1.5\,\mu m$ with a minimum loss of 5% per kilometre. One cannot obtain anything better because of the physical laws that govern the propagation of light (Maxwell laws) and because of the fundamental nature of glass. However, a recent development might revolutionize the field: the possibility of amplifying optical signals directly within the fibre without the need to extract them first. By doping the fibre with suitable elements, for example erbium, and exciting them by sending the appropriate pumping light through the fibre, one may obtain population inversion between two erbium levels in a transition that corresponds exactly to $1.5\,\mu m$. Therefore amplification of the light pulse at this wavelength may be achieved while it is travelling in the fibre. In the use of this device a piece of doped fibre is inserted between the two ends of the fibre in which the signal travels and via a suitable optical coupler both the pump light and the signal light are propagated. At the output, the remaining pump light is coupled out and the amplified signal light continues to travel in the original fibre. In this way all the amplifying stations can be eliminated. In these old stations the light was extracted, detected with a suitable detector whose signal was electronically amplified and reconverted into a light signal by a laser diode and then re-inserted in the next fibre track.

Compact discs

One of the most popular applications of lasers is the recording and reading of compact discs (CDs) which nowadays have completely substituted the old-fashioned large vinyl discs. Optical disc technology had its beginnings at Philips Research Laboratories in Eindhoven, the Netherlands, in 1969. At Sony in Japan parallel research took place. After agreement with Sony, in 1979 the two companies decided to collaborate on the development of the system and eventually the CD reached the market in 1982. In this system sound information is first registered and transformed into a series of pulses that represent the original signal (that is the signal is digitalized). Then these pulses through a complex technique are registered on the disc using a laser emitting in the ultraviolet. The laser 'writes' the pulse sequence in the form of holes made on a glass disc. Each hole has microscopic dimensions with a width of about half a thousandth of a millimetre ($0.5\,\mu m$) and a depth of $1000\,\text{Å}$, so that it is possible to register a large amount of information in the very small space available on the disc. From this first recording, a master is formed from which the copies to be sold are made usually in polycarbonate. In order to 'read' the stored information, the disc is rotated and read using the light of a semiconductor laser placed underneath. In the earliest CD players, GaAlAs diodes were used which emitted in the near infrared at $0.780\,\mu m$. Nowadays to achieve better results diodes emitting in the blue-green region are employed allowing smaller holes to be formed and therefore more information to be stored on the disc. The diode light is directed at the disc and is reflected back by the parts of the track which do not contain holes while the holes themselves do not reflect. The light back reflected is read by a suitable detector and is the reading signal. The series of light pulses and absence of pulses is then decoded with electronic methods to recover the original sound (figure 64).

Of course the system is much more complex than described. Very sophisticated optical systems must ensure that the laser light is always perfectly focused on the disc (the position for the focus does not change by more than $0.5\,\mu m$) and that the disc rotates at a rigorously constant linear speed. In the disc are also coded further signals relating to the index, that is the disc contents, the track number, the duration of the selected track and the total elapsed time. They serve to allow the selection of a desired piece of information on the disc without the need to listen to it all. Such a complicated system also has its advantages of course. The principal one is that because the disc is read by a light beam, without any direct contact, all the scratches that before were produced by the mechanical pick-up are now absent. Moreover the possibility of storing the information in a very limited space allows more information to be stored than is in reality needed and this means that if some hole is not read, this does not present a major problem since the same information is also provided by other holes. Therefore if small scratches or imperfections are produced on the

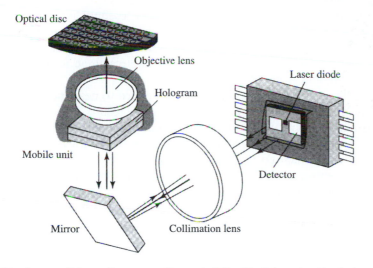

Figure 64. Layout of the reading system of an optical disc. The light emitted by the laser diode is collimated and sent to the disc for reading. The same optical system collects the reflected light and sends it to the detector.

disc, for example by touching it with hands, very often the reproduction is not affected. However, it is necessary to keep the disc well cleaned to avoid the accumulation of grease or dirt or large dust particles obscuring a large portion of holes.

Optical discs allow in general the storage of huge amounts of data. The most important application of optical discs are the so-called CD-ROMs (compact-disc read only memory) which are optical disc memories nowadays common in computers. Versions that can be written to only once are sold at the price of about 10 euros (per 10 disc pack), and rewriteable versions are also available cheaply nowadays.

Medical applications

The applications to medicine are some of the most interesting uses of lasers. Lasers have become ubiquitous in biological research and in medical practice, with dozens of major applications, most of which using standardized commercial equipment.

One of the first applications of lasers has been in ophthalmology. Photocoagulation in the form of eclipse blindness was known about in ancient times. Plato warned people never to look directly at an eclipse because it would destroy their central vision. This was forgotten by people watching the first atomic bomb explosion, some of whom were blinded by the flash. Early in the 1950s, before the laser, photocoagulators for the retina included the light from a xenon arc lamp which caused burns similar

273

to those observed after exposure of the retina to sunlight. One of the first applications was the repair of detached retinas. Xenon lamps used by doctors produced burns that could re-attach the retina onto the back of the eye. It was therefore obvious to them to develop the technique using a laser and indeed the system worked!

In photocoagulation, laser light is converted to heat energy with a rise of temperature to about 65 °C. This temperature results in the denaturing of protein and forms a coagulant. It can also destroy abnormal blood vessels, cysts, tumours, or anything abnormal in the eye. It can cause an adhesion between the choroid and the adjacent retina.

The leading cause of blindness in people in the age range 20 to 64 is vascular disease, particularly diabetic retinopathy. From 65 onwards, the leading cause of blindness is macular degeneration, which is either due to vascular insufficiency or overgrowth of blood vessels in the centre of the retina. Coagulating these blood vessels in the latter condition can stabilize these problems in many cases.

One of the pioneers in the field of laser ophthalmology was the American Francis A L'Esperance Jr, who worked at Columbia-Presbyterian Medical Center's Edward S Harkness Eye Institute in Manhattan, New York. After having experimented with the ruby laser, he convinced himself that a laser emitting in the blue-green was much more valuable. So, as soon as the argon laser—which emits exactly at these wavelengths—was invented, he managed to obtain one. He found it at the Raytheon Company which had built an argon laser for the American government with a power of 10 W that they wanted to evaluate as a weapon. It was an enormous animal more than 3.5 m long! After building it the government no longer wanted it, preferring instead another kind of laser which emitted in the red. So L'Esperance obtained the laser and, after having insured it, shipped it to his laboratory on the ninth floor of the Institute building. The laser was too big for the elevator, so he hired the best rigger in New York, who had never dropped a thing in his life. He recalls that a big crowd came out to watch the show of the laser as it was raised. However, the laser rose to almost the fifth floor and then, with a 'click click', something flew off and the crate came crashing to the ground. People were dismayed. You would have thought everybody in L'Esperance's family had been killed: people were crying and passing on their condolences.

Francis sent the crushed box back to Raytheon and amazingly the laser was still working. But the company, of course, wouldn't warranty it further. They offered to build another one. Luckily the insurance paid out and so L'Esperance could have the new laser built. Meantime Raytheon gave him a smaller laser with which Francis could start laboratory experiments which demonstrated the superiority of this laser with respect to the ruby laser. When the second laser arrived, not surprisingly he couldn't get rigging and lifting insurance. Luckily the Institute was building an extension and a

274

huge crane was sitting there. Francis went over to the crane operator, who was having lunch and asked him if he was able to raise the box with the laser inside, and for only $25 arranged for his precious laser to be deposited on the ninth floor on the terrace of his laboratory.

The laser was tested for the first time on 14 February 1968 in attempts to remove a membrane covering the pupil of the eye of a young girl. The operation was successful and the girl recovered her sight totally. Nowadays the argon laser is the standard laser for ophthalmology problems.

To treat glaucoma, when an obstruction in the pupil hinders the aqueous flux from passing between the posterior and anterior part of the eye, one may re-establish the flux by punching a small hole in the iris using a laser (iridectomy).

Lasers are used in the post-surgery treatment of cataracts and with them cataracts can be vaporized and destroyed. In the case of diabetic retinopathies, macular degeneration and vein occlusions, the laser has allowed the healing of diseases which are otherwise incurable. For many kinds of pathologies the introduction of the laser has heralded an authentic therapeutic revolution. Of enormous importance is the variety of ways in which laser techniques can help the patient. In the great majority of cases one has ambulatory treatments, made without the need of anaesthesia, recovery or particular successive treatments. Lasers are used to change the refractive power of the eye. Because this is produced by the curvature of the cornea, one may change it by sending light pulses from a laser emitting in the ultraviolet, vaporizing some corneal cells so as to decrease its curvature. In this way the focal distance is changed, obtaining the same result that normally is obtained by means of wearing spectacles.

Another application is in the removal of stones. The urinary calculousy is determined by the precipitation of the constituents of urine during its elaboration and elimination. In any stone, usually an organic matrix occurs at the centre, around which reside crystalline substances of various nature (calcium phosphate, calcium ossalate, uric acid etc.). Usually the colic is caused by the passage of the stone from the kidney to the ureter. Besides the removal by traditional surgical means, alternative techniques are employed such as pharmacological dissolution and endoscopic surgery. The crushing of urinary stones in a bloodless way is one of the most important areas of progress. The usual techniques use ultrasound, electro-hydraulic waves produced by electric discharges in the neighbouring of the stone and, beginning in 1968, laser techniques in which laser light is directed at the stone by means of an optical fibre used to vaporize and crush it. After the treatment, the fragments are of such dimensions as can travel through the urinary duct freely and be expelled spontaneously.

A number of other applications in other fields also exist. For example, in angioplasty in some cases one may use laser energy to vaporize the atherosclerotic plaque in arteries to increase blood flow. In gynaecology the laser may destroy only the area where unhealthy tissues are present, minimizing damage to the reproductive organs, decreasing blood loss and risk of infection. In oncology it is possible to heal certain tumours.

In surgery lasers are advantageously used every time the region of operation is not easily accessible and can instead be reached by means of an optical fibre that conveys the laser radiation. They are often used to cauterize blood leakages that otherwise could have deleterious effects.

Lasers are used in many therapeutic procedures and also in diagnostics. Radiation can be sent onto wounds with beneficial effects such as, for example, healing bed-sores and those produced by circulatory diseases. Other applications exist in dermatology, dentistry, orthopaedics etc. With the commercialization and rapid progress of ultra-fast lasers, new applications using such lasers are rapidly moving into the biomedical marketplace. Particularly in surgery, femtosecond pulses allow far more precise cutting than did longer pulses. Precise laser cutting is performed by photo-disruption that prevents heat damage to the surrounding tissue. We must stop here. Although we have not described everything, what we have said is enough to make us understand the fundamental importance of the use of lasers in medicine.

Mechanical work

The ability to concentrate laser light onto a very small region allows us to perform, with high speed and precision, mechanical work such as cutting, hole burning, soldering etc. on very different kinds of materials from metals to ceramics, plastics, wood etc. The possibility of controlling these operations with computers makes the new technique compatible with robotic automation.

Lasers are used to solder parts of cars, saw teeth, make holes in baby feeding bottles, cut clothes etc. The employment of lasers to cut, weld and solder sheets for the automotive industry allows us to manufacture safer vehicles and to realize a saving in terms of cost and waste production.

The laser beam can be aimed with great precision and can be sent exactly where it is needed without disturbing nearby material. It may reach small inaccessible areas. With lasers, more efficiently than with furnaces, materials that must suffer thermal treatments to strengthen them can be warmed. While a furnace must be fired 24 hours a day and is used only for a few hours, the laser uses its energy only when required. The automotive industry was the first to use the laser for the thermal treatment of axle-shafts, valves and pistons. In the treatment of surfaces the laser allows their structural properties to be changed, making them stronger, more resistant to corrosion etc.

Laser cutting is used not only in metal working, but also in the textile industry, for example to cut dresses, leather, paper or wood. The electronics industry is one of the biggest users of lasers to engrave and clean printed circuit boards and perform high speed soldering. Since 1965 the American Western Electric company first utilized the laser to precisely shape and therefore calibrate electrical resistances made with thin films.

One interesting application is for example in diamond cutting. Sometimes they present regions that do not have the same crystalline alignment as the principal structure. Because during processing, diamonds must be cut or cleaved in the direction of the principal crystal, these regions cannot be treated in a conventional way because they present knots that are very similar to the knots of a tree. The laser has demonstrated excellent ability to eliminate them.

Finally we may mention the use of laser robots as tools or sensors.

Measurement systems
Light is used in measurement for many applications. Systems involving various kinds of interferometer allow us to measure small displacements, verify the perfection of surfaces and in general to make high precision measurements.

The domain of optical non-contact measurement systems is very diverse, extending from research laboratories to chains of industrial production companies. The large variety of needs has therefore encouraged the proliferation of a great number of systems. They are used to measure the displacement of an object or its dimensions, for vibration or velocity measurements, or for the inspection of surface state. This is possible thanks to laser diodes whose light is collimated and focused on the items to be investigated. Dimensional measurements are obtained from the analysis of the light back-scattered by the object.

A non-contact optical measurement system has usually three principal parts: the measurement head, the associated electronic system and the information processing system. The principal advantage obtained by a non-contact measurement is the possibility of avoiding any deterioration of the studied surface and the fact that the system is very fast and in some cases more than 25 000 measurements per second can be accomplished.

The profile of a line on some mechanical surface is usually measured thanks to a palpatre or a stylus that passes over the surface to be measured and is moved along the line. The stylus is made of a diamond point whose contact zone with the surface to be measured has a diameter of several microns. The vertical movement of the stylus, produced by irregularities of the surface, are suitably measured and interpreted as representative of the relief of the surface (roughness). These instruments that have been improved during the years are nowadays an integral part of any micromechanics laboratory. The equivalent optical instrument uses instead a well collimated laser diode beam that is focused on the surface by means of a lens which also collects the back-reflected light from the surface and uses this information to derive the surface roughness. When the system is correctly balanced the impact point of the laser beam on the surface is exactly at the focus point of the lens and the back-reflected light is received as a perfectly collimated beam. If the surface becomes nearer to the device, the light spot on the surface is no longer perfectly focused and the lens transforms the light coming back into a diverging beam. On the contrary, if the surface moves away from the focal point, the lens brings the light back as a converging beam. A number of optical devices exist which are able to measure the degree of convergence or divergence of an optical beam. In this way the system gives an error signal that operates a servomechanism which moves the surface so that accurate focus is always maintained and so by knowing how much the servomechanism is required to move the surface, the measurement of surface relief is performed. This is also the system that enables compact-disc readers to be always exactly at the same distance from the disc, notwithstanding the irregularities of the surface or of the rotation system.

In respect of the profilometers described previously, which allow infinitesimal displacements to be measured, down to a few ångströms, the optical profilometer allows measurement without direct contact with any mechanical part. This is the only method that may be applied to the reader of a compact-disc, but it finds important applications in any situation in which the studied surface needs not to be disturbed. This is particularly important with delicate or flexible surfaces, like rubber or biological tissues, thin films or semiconductors. Other advantages are that the measuring head does not suffer any wear, that the instrument has a great lateral resolution, due to the very small size of the light spot, and allows a great speed of data acquisition.

A completely different application is the laser gyroscope, capable of measuring very small rotations. They were first made in 1963 and developed from 1965. The basic working principle was developed by the French physicist G Sagnac (1869–1928) who, in 1913, noted that light beams travelling in opposite directions along a ring could be used to sense rotation of the ring. In fact, if the point from which the beams depart moves, the two beams must travel different distances along the ring to come back to the starting point and this fact may easily be detected with a special interferometer. In 1965 laser gyroscopes were able to detect rotational velocities smaller than 5° in an hour. Such gyroscopes and other more sophisticated versions, were introduced by Boeing on its 757 and 767 jets. They are necessary components in all guidance systems.

Optical readers in supermarkets

Nowadays in every supermarket and also in most shops, a laser system reads a universal code written on the goods as an ensemble of lines (the bar code);

the obvious advantages are the acceleration of the control of prices and the reduction of errors. The laser light passes in front of the symbols, that is the series of parallel lines, black and white of different amplitude, and is partly scattered back with an intensity that is modulated in amplitude by the sequence of the lines. The back-scattered light is collected by the same device that emits the light and from its modulation the information is derived about the type of goods and its price which is then displayed on a small display next to the cash register.

Applications in construction
Very often a laser is used to align buildings. The ability of the laser beam to define by its nature a straight line was immediately appreciated. The difficulty was to make an instrument that could be easily brought onto the building site.

At the end of the 1960s helium–neon lasers were used to align the tubes of the tunnel for the Bay Area Rapid Transit System in San Francisco, California. After this, one of the first firms making helium–neon lasers, the University Laboratories of Berkeley, California, in order to overturn the initial diffidence of buyers for the new technique, painted the device a brilliant yellow colour so it looked like one of the usual pieces of equipment used in building construction.

The diverse application of lasers in the building industry started about 30 years ago with topographic measurement devices and with the employment of the laser beam as an aid to deposit draining pipes and obtain precise slopes in agriculture. The first laser levelling devices, able to project a line on a large surface, were developed 20 years ago. British Rail was one of the first companies to use the laser guidance for the building of plants, with conspicuous economic advantages. Lasers are employed in road construction by using laser guidance for the heavy machines that prepare and put large surface coatings on the ground. A laser system employs the design data provided by the engineer for the guidance of the machine that places on the ground a composite surface such as, for example, a road curve that skirts a hill. The system is used for the construction of roads, railways, motorways and other plane surfaces with tolerances around 2 cm if hand controlled, which may decrease down to 5 mm with automatic control. Curves can be accurately controlled both vertically and horizontally.

The atmosphere
Lasers allow us to study the atmosphere. The device that makes this study possible is called the lidar. In a lidar, similarly to a radar, the light back-scattered by molecules or particles (aerosol, water droplets etc.) in the atmosphere is measured. The simplest system detects the light coming back,

identifying the presence of aerosol, water droplets, clouds, smoke particles etc. In this way concentration profiles as a function of altitude can be obtained and it is possible to follow their evolution with time. It is also possible to measure the presence and concentration of a particular kind of gas using a lidar emitting at two different wavelengths: one which is absorbed by the gas under study and the other which is not absorbed. Of course the two return signals will be different—the non-absorbed signal being stronger—and from their difference, the particular gas species can be localized and their concentration measured. These lidar are called DIAL (Differential Absorption Lidar). DIALs have been built in several towns and used to measure polluting substances such as pesticides or smoke or gases such as SO_2, O_3, NO_2 and NO originating from industrial plants.

Urban air pollution typically has many sources. These commonly include fuel combustion, industrial emissions, and emissions from fires, all of which mix with natural air. Tracing and identifying all of these contributions is necessary for understanding and potentially mitigating pollution sources.

In Europe several measurement campaigns have been carried out to measure urban pollution. For example in Lyon, France, a DIAL was located near the Fourière cathedral to measure the NO emission from the traffic in the town centre; the highest concentration was found to be in the old part of the town where the traffic is slow and ventilation is hampered. The results suggested that a pedestrian zone should be created in this area. In Stuttgart, Germany, which is surrounded by several hills, it is possible in certain conditions to observe trapping of polluted air due to traffic, principally NO at a height of 450 m. In Berlin the SO_2 missions were much higher in the eastern part of the town due to the poor quality of the carbon used for heating, while the high pollution concentration at ground level in the Alexander Platz was swept away by the south-west wind. The SO_2 concentration that contributes significantly to the production of acid rain and is produced by the smoke produced by home heating, by traffic and by power stations has been measured at Lipsia, Germany. The results showed that domestic heating and traffic are the major problems.

These measurements have shown that topographical and meteorological conditions often play a more important role than the intensity of the emission of the pollutant: narrow streets with low traffic generate higher concentrations than large and well ventilated streets even with intense traffic.

In summer 1994, a campaign was launched in Athens, organized by the European Community, to clarify the photochemical and meteorological processes which generate the summer fog in the Greek basin. It was discovered that the role of the chemical reactions between NO, generated by traffic, and NO_2 and ozone was essential.

An important application is the use of lidar to measure the concentration of ozone (O_3) in the stratosphere over the Antarctic and Arctic. The quantity and distribution of ozone in the atmosphere may have important implications for pollution, for chemical processes in the troposphere and for those related to the climate, besides the screening of ultraviolet solar radiation and the chemical and thermal equilibrium of the stratosphere. The ozone present in the high atmosphere, in fact, absorbs solar ultraviolet radiation which has adverse biological effects and thus prevents it reaching the ground. It is well known that in the past few years a consistent decrease in the thickness of the ozone layer in the atmosphere over the poles has been observed (the so-called 'ozone hole'). This hole is produced by a series of reactions with certain volatile products used in several industrial applications (for example in the refrigeration industry or in aerosol sprays). The vertical distribution of ozone—which dictates the spatial and temporal development of the 'hole'—has been measured with several lidar.

The lidar may be used for measurements of wind velocity—necessary for meteorological purposes and for the development of climate models, the speed of aerosols, smokes etc. In this case one uses an effect, known as the Doppler Effect, which consists of a small change in the frequency of the light emitted or reflected by a moving body. By measuring the small change of the frequency of the reflected light with suitable methods, one may obtain information on the velocity. The instruments used are called Doppler lidar.

With lidar convection phenomena around clouds can be studied. Clouds reflect and irradiate heat and are important to warm or cool the atmosphere, but nobody yet knows exactly how to describe their influence. One may also study the water vapour that plays a role in the formation of storms. By measuring the changes in the spectra of the oxygen molecules, atmospheric pressure and temperature are derived.

Marine measurements can also be performed using lidar to measure the quantity of chlorophyll and phytoplankton in the ocean: the importance of this last measurement is evident if we observe that phytoplankton very probably supplies two thirds of the total amount of oxygen of our planet. Loss of petroleum and other pollutants from ships can also be measured in the water as well as water temperature or salinity. Combustion phenomena can be studied. The purpose of several research programmes is to develop diagnostic techniques for combustion processes and design measurement systems for industrial plants. Gas temperature, concentration and velocity can be measured, together with the dimensions of smoke particles.

For all these measurements all kinds of lasers are used, according to the particular application: ruby, neodymium, dyes, diode lasers and others, systems with a huge cost variation, either in fixed position or as portable units.

Adaptive optics

We wish now to describe certain applications which at first sight may appear as ideas for science fiction. One of these is the so-called adaptive optics.

Adaptive optics improves the quality of the image provided by large telescopes, by compensating for the aberrations induced by the atmosphere, that is the distortion that it produces on light beams. These distortions are easily seen when one watches, for example, a distant landscape at sunset on a hot, still day when the image appears to tremble. Adaptive optics compensates for these irregularities and sometimes is defined as 'the technology that stops stars twinkling' a definition that may generate the horrified reaction 'It is terrible and should not be allowed!'

Let us see now what happens. Stars are so distant that their light arrives at the Earth as a wave whose surface is a plane (planar wave front). In theory, a telescope equipped with perfect optics should concentrate this light into a small bright circular spot whose dimension is limited only by the diffraction phenomenon, that is by the effect of the dimension of the principal lens or mirror of the telescope on the arriving wave. Two nearby stars may be seen distinctly separated if the angle under which they are seen by the telescope is larger than a minimum value at which the two bright spots produced by each of them merge into one. This minimum angle is called angular resolution. Lord Rayleigh gave a criterion to define this quantity. A telescope's angular resolution in arc seconds is equal to a constant times the wavelength of light divided by the telescope's aperture. The Hubble Space Telescope with a diameter of 2.4 m, in orbit around the Earth, should have an angular resolution of about 0.05 arc seconds, that is it may distinguish between details as close together as 0.05 arc seconds. On the ground, a similar 2.4 m telescope, due to the distortion introduced by the atmosphere, has angular resolution that is 20 times larger (about 1 arc second).

Telescopes are built with large apertures of high quality. Gigantic light collectors may in fact detect and measure properties of objects that are very faint, because with their large aperture they are able to collect a great number of photons emitted by the object. Moreover, telescopes with a high resolving power are able to discern more details from the observed objects. Unfortunately small temperature fluctuations in the atmosphere induce fluctuations of the refractive index of the air and so different parts of an initially plane wave front are affected in different ways and the net result is that the images in the telescope are blurred. These are the aberrations we were speaking of earlier. The image of the disc of a star produced by a 4 m telescope on the ground is typically 40 times larger than the optimal value allowed by diffraction theory. The dimension on which a wave front remains sufficiently plane determines the enlargement. Technically it is referred to as the coherence diameter of the atmosphere and generally it is 10–20 cm in a good observation spot. The fact that precious photons coming from the studied object are sparse over an area typically 40 times greater than diffraction allows, implies that the intensity of the observed image is reduced by a factor of 40 squared. Therefore, even though a large telescope can collect more photons, if its aperture is larger than the coherence diameter of the atmosphere, nothing is gained in resolution. Critics could interpret this as a licence to define the world's largest telescopes as over-priced!

Isaac Newton in *Opticks* in 1730 wrote:

'If the Theory of making Telescopes could at length be fully brought into Practice, yet there would be certain Bounds beyond which Telescopes could not perform. For the Air through which we look upon the Stars, is in perpetual Tremor; as may be seen by the tremulous Motion of Shadows cast from high Towers, and by the twinkling of the fix'd Stars. But these Stars do not twinkle when viewed through Telescopes which have large apertures. For the Rays of Light which pass through divers parts of the aperture, tremble each of them apart, and by means of their various and sometimes contrary Tremors, fall at one and the same time upon different points of the bottom of the Eye, and their trembling Motions are too quick and confused to be perceived severally. And all these illuminated Points constitute one broad lucid Point, composed of those many trembling Points confusedly and insensibly mixed with one another by very short and swift Tremors, and thereby cause the Star to appear broader than it is, and without any trembling of the whole. Long Telescopes may cause Objets to appear brighter and larger than short ones can do, but they cannot be so formed as to take away that confusion of the Rays which arises from the Tremors of the Atmosphere. The only Remedy is a most serene and quiet Air, such as may perhaps be found on the tops of the highest Mountains above the grosser Clouds.'

Obviously some system is necessary to rectify the distortion effects of the atmosphere, known since Newton's time. This system is adaptive optics. Historically one may quote as the first use of adaptive optic the destruction by Archimedes in 215 BC of the Roman fleet. As the Roman fleet approached Syracuse, soldiers lined up so that they could focus sunlight on the sides of the ships using their shields as mirrors. In this way hundreds of beams were directed towards a small area on the side of a ship and the resulting intensity was enough to ignite the ship and defeat the attackers. This ingenious idea passed into legend as the 'burning mirror' of Archimedes.

In 1953, H.W. Babcock, then the director of the astronomical observatory of Mount Wilson in California, proposed, using a deformable optical element driven by a wave front sensor, to compensate for atmospheric distortions that affected telescope images. This appears to be the earliest suggestion of using adaptive optics.

Most of the pioneering work in adaptive optics was carried out by the American military in the 1970s and 1980s. The military were interested in its application to the propagation of lasers in the atmosphere and for an improved positioning of satellites and guidance of missiles in flight. The research was strictly classified. The first adaptive optics system—still used—was installed in 1982 by the Air Force at the optical station of Maui in Hawaii.

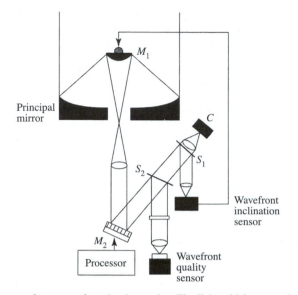

Figure 65. Layout of a system for adaptive optics. The light which enters the telescope first encounters the mobile mirror M_1 which corrects the wavefront inclination. The residual aberrations are then corrected by the deformable mirror M_2 and the cleaned wave is eventually sent to the detector C. Part of the light collected by the tilted mirrors S_1 and S_2 is employed to direct the two mirrors M_1 and M_2.

The astronomical community started to develop experimental adaptive optics systems early in the 1980s, when the majority of military work was still classified and the two research programmes—one involving astronomers and the other the military—progressed in parallel without any exchange of information. Initially there was great scepticism about whether the technique was useful and it was difficult to obtain financial support. In 1991 the situation changed. Most material was declassified and telescopes started to produce images more clearly focused as a result of adaptive optics. Since then the military and the academic community have worked together.

Figure 65 shows the general layout of a telescope that uses adaptive optics. The wavefront sensor samples an incoming wave front in order to measure the local deformations. The processor translates this information into a signal that immediately must be used to correct the wavefront. The correction, in real time, must produce an aberration equal and opposite to that produced by the atmosphere. The operation has to be repeated with the same speed with which the atmosphere produces the changes, typically between 10 and 1000 times per second. In conventional systems the correction is generally made with a deformable mirror made by a thin membrane controlled by a series of piezoelectric actuators affixed to the mirror's back.

Information on the distortion of the wavefront can be derived from the target itself if it is a point source (a star) and fairly bright—brighter than the sixth magnitude at visible length (the faintest stars discernible by the naked eye). Many objects of interest to astronomers, however, are not point sources but extended objects (such as planets or nebulae) and most are thousands of times fainter than the sixth magnitude. In these cases one may use a nearby star to provide the reference wavefront, but its light should travel through the same portion of atmosphere as the light of the object under study. This imposes the condition that the star should stay within an angular diameter of 2 arc seconds which is in reality a very small portion of sky in which it is difficult to find a star bright enough. This leaves just one alternative: to create an artificial guide star or beacon brighter than the sixth magnitude. At this point the laser enters the field. An artificial source is created by illuminating with a powerful laser a region in the high atmosphere where a substance exists that when illuminated re-emits light. Sodium, which is in high concentration between 80 and 100 km in the upper atmosphere, can be used. To excite sodium—the D-line, of course—a laser at 5890 Å is used. Systems involving laser guiding stars have been built for example near the Philips Starfire Optical Range Laboratory at Albuquerque in New Mexico, at the Calar Alto Observatory in Spain and at the Lick Observatory in California.

Shortly, astronomers will be able to measure the diameters of stars brighter than the tenth magnitude; observe sunspots on their surface and measure changes in their position with sufficient precision to ascertain whether they are encircled by planets. Thus, the enormous progress achieved allows us to believe that with this technique it could even be possible to see the presence of planets near distant stars. The planets must be identified against the background of the light scattered by the star around which they revolve, which is about a factor of 10^9 times brighter than them. For the detection of planets, however, we are operating in favourable conditions to make accurate corrections because in this case the star around which they are orbiting can be used as the reference source. The next generation of terrestrial telescopes will thus have the potential to detect planets orbiting around the stars nearer to us.

Spectroscopy

If we turn now to more basic applications, we need to mention spectroscopy. When dye lasers were invented and it became apparent that their wavelength could be varied over some given range, one understood immediately that they were an ideal source for spectroscopy. These lasers allowed new levels of sensitivity and resolution. The explosion of lasers in spectroscopy occurred in the 1970s. A laser may, for example, vaporize a minuscule piece of a sample to be analysed and allow extremely precise microanalysis. A number of very skilled researchers have used lasers for spectroscopy, among them Shawlow who earned the 1981 Nobel prize for physics specifically for the development of laser spectroscopy.

It has since been shown that single atoms can be detected, controlled and manipulated. In one experiment, a single caesium atom has been detected and identified in a vessel containing 10^{18} other atoms. Atoms can be cooled to the low temperature of only one millionth of a degree higher than absolute zero, and by means of fast pulses the details of events occurring within molecular reactions can be studied in the time an electron takes to turn around a nucleus. In 1997 the Nobel prize was earned by the physicists C Cohen-Tannouji, S Chu and W D Philips for their contribution to the development of methods to cool and trap atoms using the laser, recognizing their ability to use spectroscopic methods to achieve their results.

Geophysics

Geophysicists use satellites, which reflect back laser light to measure the movements of the Earth's crust. By measuring the time taken by the laser pulse to go and come back from the satellite, one may measure with great precision the distance between the laser and the satellite. If the satellite turns on a fixed orbit so that its distance from the Earth does not change in time, this method allows small displacements of the laser to be monitored from which measurements of continental drift may be derived.

Continents are floating over the warm internal layer of the Earth as plates of terrestrial crust. These plates by colliding with each other may provoke earthquakes, induce the emergence of islands or the explosion of volcanic eruptions. The measurement of continental drift is therefore of great importance. The LAGEOS (Laser Geodynamics Satellite) satellite programme in the 1970s provided proof of continental drift. Measurements continue today with a second geodynamic satellite. For example, measurements are performed along the fault line existing in California to measure small displacements in an attempt to predict earthquakes before they happen.

In the same way one may see how the Earth turns around its axis and changes its shape.

The laser and the Moon

Bell Laboratories used one of the first lasers to study the roughness of the Moon's surface. During the Apollo 11 mission that landed on the Moon on 21 July 1969, the astronauts positioned on the surface two mirrors that would reflect laser light sent from Earth. A group of astronomers at the Lick Observatory in California sent a powerful beam from a ruby laser to the Moon and succeeded in receiving the reflection, thus measuring the

Earth–Moon distance with a precision greatly superior to that obtained by the usual astronomical observations.

A laser altimeter has been used in the Mars Orbiter Laser Altimeter (MOLA) to give a global three-dimensional view of Mars.

Gravitational waves

In 1918 Einstein predicted that moving masses produce gravitational waves propagating at light velocity. Unfortunately the amplitude of the gravitational radiation emitted by any source that could be built in a laboratory is too small to be detected. Astrophysical phenomena which imply the motion of very large masses at relativistic velocities may on the contrary produce enough gravitational radiation to be detected. Indirect evidence of their existence has been found, and it was enough to earn Alan Russell Hulse (1950–) and Joseph Hooton Taylor (1941–) the 1993 Nobel prize for physics, but definitive direct evidence is still lacking. Gravitational waves arise from acceleration of masses much like electromagnetic waves are radiated from the acceleration of charges. They affect masses stretching an object in one direction while compressing it in a perpendicular direction.

When a gravitational waves passes, it may set a mass into oscillatory motion up and down like the ocean waves. To detect gravitational waves it is necessary to measure this motion.

In principle the distortion produced by a gravitational wave may be measured with a large mechanically insulated cylinder which resonates mechanically at the frequency of the gravitational wave. Suitable sensors convert the vibrations into electrical signals which are measured. The first detector with a resonant cylinder was designed in the late 1950s and built early in the 1960s by Joseph Weber of whom we already spoke when describing the maser. Weber built an aluminium cylinder of several tons in weight which was resonant at a frequency of about 1 kHz. He claimed positive results but nobody could duplicate them. Subsequently other detectors of this kind were built in a number of institutes around the world. The better of these devices is able to detect a displacement of one part in 10^{12}. But this has still proved insufficient to detect gravitational waves unless they are produced very near to us and by extremely violent events.

An alternative way to detect gravitational waves consists of measuring the time light takes to travel between two mirrors suspended as two heavy pendulums, which are the masses that may be set into oscillation by the gravitational wave. The method involves comparing the transit times of two laser beams that travel at right angles in a Michelson interferometer (like that used to measure the speed of light in a vacuum). The gravitational wave would compress one path making it shorter and stretch the other. Figure 66 shows a possible layout. Experiments started in the 1970s. If the interferometer arm is 4 km long, a typical gravitational wave changes its

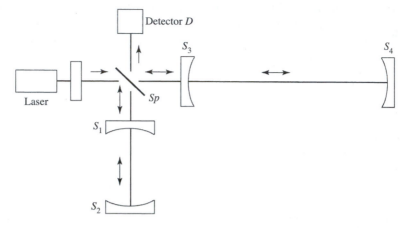

Figure 66. The interferometer to detect gravitational waves is made by four mirrors S_1, S_2, S_3, S_4 suspended on vibration-isolated weighted pendulums to create two mutually perpendicular light paths $S_3 S_4$ and $S_1 S_2$. A laser beam is split into two beams by the semitransparent mirror Sp and travels back and forth many times between the two pairs of mirrors before recombining in the detector D. If a gravitational wave passes by, the pendulums holding the mirrors are expected to move apart in one arm and together in the other by a tiny fraction of the laser light wavelength. The movement would shift the relative phase of the two halves of the split laser beam changing the interference conditions on the detector and registering as a detection of a gravitational wave.

length by less than 10^{-14} m, that is about one thousandth the dimension of an atomic nucleus. In the interferometer light travels many times back and forth between a fixed mirror and a mirror attached to the pendulum so that the length difference sums as many times as there are reflections. Interferometers of this kind have been built in several places in the world. Weber understood already in the 1970s that a laser interferometer would have been more sensible than the cylinder approach and the same thing was proposed, independently, by the Russian Michail Gerstenstein and V I Pustovoit of Moscow University and Ranier Weiss of MIT. The first interferometer was built in 1978, and in 1983 an interferometer 40 m long was installed at the Californian Institute of Technology. Similar interferometers exist nowadays in Italy, Germany and Japan. Recently an even more powerful device has been designed with an interferometer 4 km long in which the travelling light is protected within a tunnel. Two versions of this interferometer have been realized at Hanford, in Washington State, and one at Livingston in Louisiana. These interferometers dubbed LIGO (Laser Interferometer Gravitational-wave Observatory) should have a sensitivity of one part in 10^{15} which will possibly increase by a factor of 100. Work has been progressing on this project since August 2002.

In Italy this research is very active. An Italian–French project named VIRGO has been built in Cascina near Pisa with arms 3 km long. It was officially opened in July 2003. Astrophysicists expect that LIGO and VIRGO will be able eventually to detect gravitational waves produced from highly relativistic events such as a collision between two black holes because until now there has been no certain direct detection of them.

German and British physicists have built a 600 m device called GEO600 near Hanover and a smaller device 30 m long (TAMA detector) is operating near Tokyo.

Ultra-short pulsed lasers

With special techniques it is possible to build pulsed lasers emitting short pulses shorter than 10^{-15} s (femtosecond). These times are so short that they can be compared with the revolution time of one electron in its orbit around the nucleus of an atom. With these pulses one may investigate chemical, biological, physical phenomena etc. which last only for a short while in a space sometimes corresponding to the dimensions of a few molecules. Using such pulses, a group of chemists, for example, has studied the behaviour of photochromatic glasses. These materials are familiar to anybody who wears the type of sunglasses that change in colour according to the light intensity. The group has shown that they change in colour by modifying their molecular structure in a period of a few picoseconds. Ultra short pulses have found industrial applications also in metal working.

Nonlinear optics

Before the laser, transparent optical materials were considered essentially as passive objects not influenced by the passing light. The high power of laser beams has made it possible, for the first time, to observe that the presence of light may itself influence the medium. Intense light may, for example, change the refractive index of a medium or its absorption. When this occurs, the light itself is affected by the change so that the final result is no longer independent of the light intensity but has a complex dependence on it. In such cases one speaks of nonlinear optics.

The nonlinear response of the material may convert the laser light into new colours. This possibility is extremely important in practice because, even though a great number of lasers exist, any kind of laser usually generates only one or a few frequencies and only a few lasers are commercially available. The need to have new wavelengths and to change them has thus enhanced interest in exploring the possibilities offered by nonlinear optics.

The observation that intense light may create changes that act on the light itself, was initially seen as a problem in the transmission of powerful laser beams through optical materials. According to the properties of the material, the light could self-focus or self-defocus itself, destroying the material or destroying itself. Later these properties have been exploited in information devices, to build light switches, couplers and to process information. The nonlinear response may be extremely fast, typically of the order of a picosecond.

The change of the refractive index induced by the light itself may facilitate the production of particular light pulses, called solitons. In optical fibres, temporal solitons are light pulses that retain in themselves a constant time duration counteracting the dispersion phenomenon which would act to broaden them. A light pulse is formed by the superposition of light rays of different colours that due to dispersion travel at different velocities so that after some distance the pulse has broadened. If the pulse is sufficiently bright, the induced nonlinearity compensates exactly this effect and the pulse may travel in the fibre for thousands of kilometres without changing its temporal profile.

Another kind of soliton also exists, the so-called spatial soliton, in which the nonlinearity compensates exactly the diffraction effect that produces a transversal widening of the pulse during its propagation. In this way a spatial soliton may propagate over great distances without changing its spatial dimension.

The properties of solitons and their mutual interactions make these pulses particularly suitable to build devices such as light switches and couplers and to be used for transmission in optical fibres. Solitons in the future may also form the basic element for optical computers.

Quantum cryptography

We wish now to consider one of the most curious and intriguing applications of lasers, quantum optics and quantum mechanics: so-called quantum cryptography. It is one of the fantastic applications made possible by lasers and by the laws of quantum mechanics.

Quantum cryptography is a new technique for the secure transmission of information. In contrast to conventional methods of cryptography, the secrecy of the information transmitted with quantum cryptography is secured by the laws of physics. Cryptography has had a long and distinguished history in military and diplomatic use dating back to the ancient Greeks. Nowadays, secure communications are becoming increasingly important for commercial applications. In addition to its practical potential applications, it illustrates several interesting aspects of quantum optics, including the role of Heisenberg's Uncertainty Principle in optical measurements and two-photon interferometry.

The first methods of cryptography employed a secret key to encrypt messages before transmission and to decode them upon arrival. The security of these methods was often compromised by the unauthorized spreading of the key or by an analysis sufficient to identify the key by trial and error and thus break the code. Most modern methods do not use a secret key but rely on the mathematical difficulty that recovering the message presents as a result of having to search through all the possible combinations to find the correct one. In any case the security of these methods can be compromised by unexpected advances in the mathematical technologies of deciphering or in computer power.

Quantum cryptography uses a secret key to encrypt and decrypt information that is transmitted on a public channel, but the key is not transmitted in the usual way. One method of quantum cryptography establishes identical keys in two distant locations without transmitting any information. Although this may seem impossible from the point of view of classical physics, it is made possible by the non-local properties of two photon interferometers. In another method, on the other hand, the key is transmitted in the form of single photons and the uncertainty principle of quantum mechanics secures the impossibility that unauthorized personnel can intercept the message. All methods of quantum cryptography are based on the principle that in quantum mechanics any measurement perturbs a system in an unpredictable way. To explain the details of how this fascinating application works is not easy. We will limit ourselves to providing some ideas of the case in which the so-called method of two photon interferometry is used.

Let us consider figure 67. Two people, Alice and Bob, at a great distance from each other, have two identical interferometers made by two totally reflecting and two partially reflecting mirrors, as shown in the figure. One photon which arrives on any one of the two interferometers, for example the left one, according to quantum mechanics has two possibilities: either it travels from S_1' to S_2' directly, or it follows the path S_1', S_3', S_4', S_2'. If the two paths are very different from each other no interference phenomena occur and therefore the photon in the first case exits in the direction $2A$, while in the second case it comes out in the direction $2B$. The same occurs for a photon arriving on the other interferometer. The possible outcomes A and B are labelled as $1A$ and $1B$ for the interferometer on the

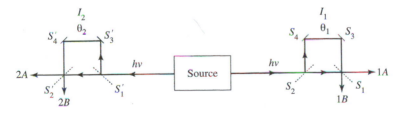

Figure 67. Two-photon interferometric method. The two interferometers I_1 and I_2 are made by the four (totally reflecting) mirrors S_4, S_3, S_4', S_3' and the four partially reflecting mirrors S_1, S_2, S_1', S_2'. The outputs $1A$ and $2A$ represent, for example, the bit 0, while the outputs $1B$ and $2B$ represent the bit 1.

right and 2*A* and 2*B* for the interferometer on the left so as to distinguish them.

Now for the best bit! One of the possibilities offered by nonlinear optics is the production of new colours that are obtained because in the nonlinear material two photons which have some frequency—i.e. some energy—merge into one photon that has an energy which is the sum of the energies of the single photons and therefore has a frequency that is the sum of the two frequencies. If the two photons have the same frequency, the new photon has a doubled frequency. This phenomenon is known as the production of second harmonics. If the two photons have different frequencies one speaks of a parametric effect. It is possible also to obtain another inverse process in which the photon in the nonlinear interaction breaks into two photons which each have a frequency exactly one half the initial photon frequency. This process is called (degenerate) down-conversion. The laws of the process guarantee that the two photons are emitted at the same time, even if quantum mechanics prevents by the uncertainty principle knowing at which time they are issued (as their energy is perfectly known).

Let us now assume that the source that emits these photons is located midway between the two observers. The process may progress such that one photon is sent to the right interferometer and the other to the left. If the detectors at right and left are adjusted in such a way as to give a signal only when they both receive a photon, the circumstance that the two photons were emitted simultaneously implies that if a photon is detected in 1*A*, the other must be detected in 2*A*, and vice-versa, if the first one is detected in 1*B*, the second must be detected in 2*B*. No signal has been exchanged between Alice and Bob, but if Alice detects a photon in 1*A* she knows that Bob has also detected a photon in 2*A*. In this way the two observers have the same signal without having exchanged any information. If now the photon detected in *A* is taken to represent an information bit '0' and the photon detected in *B* represents the bit '1', by observing the casual sequence of photons emitted by the source, the two observers both receive the same casual sequence of 0 and 1 bits that constitutes the secret key with which to transmit and read a message. No information has been sent between Alice and Bob to establish the secret key because the output of the interferometer is undetermined until the measurement has been done. At that moment quantum mechanics requires that, if the right interferometer measures a photon through 1*A*, the left interferometer must detect it through 2*A*. If someone wishing to detect the photons inserts himself along the line of transmission from the source to one of the interferometers, then clearly that intercepted photon is not detected at one of the interferometers and therefore the other interferometer does not detect anything either because there is no coincidence of signals. That photon simply does not participate in the establishment of the secret key common to the two observers.

Systems of cryptography as the one we have described, or founded on different kinds of experiments, have been demonstrated experimentally in recent years and look very promising.

Atom trapping

The 1997 Nobel prize was awarded for the development of cooling and trapping methods by means of lasers to Steven Chu (1948–) of the University of Stanford, Claude Cohen-Tannouji (1933–) of the College de France and the Ecole Normale Supérieure of Paris and William Phillips (1948–) of the National Institute of Standards and Technology, Maryland.

The cooling and trapping of atoms by means of light are two distinct processes that are, however, connected. Because optical traps for neutral atoms are generally shallow, it is necessary to cool the atoms below 1 K before thinking of trapping them. The cooling of an atomic gas by a laser was proposed in 1975 by Theodor Haensch and Arthur Shawlow at the University of Stanford. In the same year, David Wineland and Hans Dehmelt at the University of Washington, Seattle, suggested a similar scheme to cool ions. The work on ions earned Dehmelt (1922–) and Wolfgang Paul (1913–1993) of the University of Bonn a share of the 1989 Nobel prize 'for the development of the ion trap technique'. The other recipient was N F Ramsey.

The principle of laser cooling is the transfer of momentum by a photon to an atom. The atom absorbing the photon receives a push in the direction in which the photon was travelling. In the subsequent re-emission of a photon, the excited atom recoils. If the emission is spontaneous the direction of the re-emitted photon is random. A series of absorptions and re-emissions transfer momentum to the atom in the direction of the laser beam, while the recoils average to zero. The result is that an atom which is propagating against the light beam is slowed, very much like a cyclist riding against the wind.

In the 1960s, Phillips and collaborators used this principle to slow a beam of sodium atoms and in 1985 trapped the cooled beam by means of a magnetic field.

In 1985 Chu and collaborators, using six laser beams, formed beam pairs orthogonal to each other and succeeded in cooling an atom gas in which every atom moves randomly in any direction.

Three years later Cohen-Tannouji discovered a way to cool atoms at temperatures impossible by these simple methods using quantum interference processes by means of counter propagating laser beams. In 1995 he succeeded in cooling a helium atom gas in space to the fantastic temperature of only 4 millionths of a degree above the temperature of absolute zero.

The cooling and trapping techniques of neutral atoms have been essential in the demonstration of Bose–Einstein condensation and may allow the production of clocks with an unimaginable precision and ultra-precise methods for the measurement of gravity etc.

Bose–Einstein condensation

Certainly one of the most spectacular results of modern physics has been achieved with the direct experimental proof, obtained in 1995, of Bose–Einstein condensation. Einstein had predicted, in 1924, the existence of a special state of matter in which atoms with certain properties, the so-called bosons— particles with a total spin that is an integer multiple of h—may be forced to stay in a state in which they all have identical quantum properties. In 1995 Eric Cornell (1962–) and Carl Wieman (1951–) at the National Institute of Standards and Technology and of the University of Colorado succeeded in cooling rubidium atoms with a laser beam and confining them in magnetic traps. Further cooling was then obtained using a method called evaporative cooling, working in the same way with which a cup of tea cools, that is allowing the hotter atoms to escape.

At the very low temperature achieved, the atoms in the new state instead of travelling in all directions, as occurs in an ordinary gas, move altogether at the same velocity and in a same direction. They have lost their identity and now become a single collective unit, their organized configuration giving rise to strange properties. The Bose–Einstein condensate was produced in a cloud of rubidium-87 atoms which were cooled down to about 170 nK and in the most completely condensed samples about 2000 atoms were in a single quantum state for longer than 15 seconds. At MIT Wolfgang Ketterle (1957–) and his group formed a sodium-23 condensate having about one hundred times more atoms than JILA. Cornell, Ketterle and Wieman were awarded the 2001 Nobel physics prize 'for the achievement of Bose–Einstein condensation in dilute gases of alkali atoms and for early fundamental studies of the properties of the condensate'. With the Bose– Einstein condensate it is possible to explore certain aspects of quantum mechanics and maybe understand superconductivity: the property of some materials to carry an electric current without any resistance. Also the origin of the Universe in certain theories is connected to Bose–Einstein condensation.

The behaviour of condensed atoms is comparable with that of normal atoms as laser light can be compared with the light from a lamp. With laser light, all photons are in phase, a property that makes the laser beams powerful and able to be focused to a very small spot. In the same way the atoms of a Bose–Einstein condensate are all in phase and physicists are working to make them behave as an 'atom laser'. Such an atomic beam allows manipulations and measurements to be made on incredibly small scales. In an atomic laser atoms could be moved one at a time. Such atomic lasers could be used to lay down atoms on a substrate with extraordinary precision, substituting photolithography. It would be possible to built atomic interferometers that, because the wavelengths of atoms are much smaller than those of light, could be used for making measurements far more precise than can be made with laser interferometry, improve atomic clocks, obtain nonlinear interactions similar to the optic case, and so on.

We could present many other applications and future perspectives of lasers, but we hope that what we have said is sufficient to make it clear what extraordinary potential laser devices offer to modern society.

BIBLIOGRAPHY

J Bennett, M Cooper, M Hunter and L Jardine, *London's Leonardo: the Life and Work of Robert Hooke*, Oxford University Press 2003

David Wilson, *Rutherford: Simple Genius*, MIT Press, Cambridge, Mass. 1984

Louis Brown, *A Radar History of World War II: Technical and Military Imperatives*, Institute of Physics Publishing, Bristol 2000

D Marconi, *My Father Marconi*, Guernica Editions Inc, Toronto 1996

Nick Taylor, *Laser*, Citadel Press, Kensington Publ. Corp. 2000

C H Townes, *How the Laser Happened*, Oxford University Press, New York 1999

J Mehra and H Rechenberg, *The Historical Development of Quantum Theory, vol I*. Springer-Verlag, New York 1982

M Jammer, *The Conceptual Development of Quantum Mechanics*, McGraw-Hill Book Co., New York 1966

O Darrigol, *From c-numbers to q-numbers*, University California Press, Berkeley 1992

A Pais, *Einstein Lived Here*, Oxford University Press 1994

A Pais, *Subtle is the Lord: Science and Life of Albert Einstein*, Oxford University Press 1982

A Pais, *Neils Bohr's Times: In Physics, Philosophy and Politics*, Oxford University Press 1991

M Bertolotti, *Masers and Lasers: An Historical Approach*, Adam Hilger, Bristol 1983

J L Bromberg, *The Laser in America 1950–1970*, MIT Press, Cambridge, Mass. 1991

Carlo Cercignani, *Ludwig Boltzmann: the Man who Trusted Atoms*, Oxford University Press, New York 1998

T S Kuhn, *Black Body Theory and Quantum Discontinuity 1894–1912*, Oxford University Press 1978

R S Westfall, *The Life of Isaac Newton*, Cambridge University Press, New York 1993

D C Lindberg, *Theories of Vision from al-Kindt to Kepler*, Chicago University Press 1976

Bibliography

Peter Harman, *The Natural Philosophy of James Clerk Maxwell*, Cambridge University Press, 1998

E Whittaker, *A History of the Theories of Aether and Electricity*, Harper and Brothers, New York 1960

H von Kroge, *GEMA: Birthplace of German Radar and Sonar*, Institute of Physics Publishing, Bristol 2000

INDEX